Quantum-mechanical tunnelling in biological systems

Quantum-mechanical tunnelling in biological systems

Don DeVault

Visiting Associate Professor, Physiology and Biophysics Department, University of Illinois at Urbana–Champaign

SECOND EDITION

The right of the University of Cambridge to print and sell all manner of books was granted by Henry VIII in 1534. The University has printed and published continuously since 1584.

Cambridge University Press

Cambridge

London New York New Rochelle

Melbourne Sydney

Published by the Press Syndicate of the University of Cambridge
The Pitt Building, Trumpington Street, Cambridge CB2 1RP
32 East 57th Street, New York, NY 10022, USA
296 Beaconsfield Parade, Middle Park, Melbourne 3206, Australia

First published in *Quarterly Reviews of Biophysics*, vol. 13, no. 4, 1980
Second edition 1984

Printed in Great Britain at the University Press, Cambridge

Library of Congress catalogue card number: 83-15445

British Library cataloguing in publication data

DeVault, Don
Quantum-mechanical tunnelling in biological
systems

1. Chemical reactions, Conditions and laws of
2. Electrons
I. Title
539.7'2112 QD501

ISBN 0 521 24904 X

To the memory of my parents
Ralph Pulliam DeVault
and
Ruth (Tenney) DeVault
and to the memory
of my father in science
Professor Arthur Amos Noyes
and
to my wife
Roberta Jeanette (Baird) DeVault
who has to put up with me now

Contents

Preface

This book is a revision with extensions and updating of the paper of the same title appearing in *Quarterly Review of Biophysics* **13** (1980), 387–564, and is a response to the demand for reprints, the supply of which was long ago exhausted. The goal was and continues to be to clarify for experimentalists, especially in the biological fields, the mechanisms and current theories concerning electron-transfer reactions, besides reviewing the pertinent experimental data.

Electron transfer, especially over the distances common in biological systems usually involves quantum-mechanical tunnelling of the electron and often of the atoms and molecules as well, so the subject of nuclear tunnelling with or without electron tunnelling is presented also. In all cases there is interaction with the vibrations of the molecules involved, and the treatment of these 'phonon' phenomena occupy most of the theory.

The review of tunnelling in physics and in chemistry is intended only to give examples and provide background for the biological cases and so is not at all comprehensive. The review of the biological data is intended to be more complete but no attempt has been made to be exhaustive. The treatment of theory is intended to give the reader the benefit of my own struggles with the theory which is not easy for the non-specialist to understand in the forms found in the original literature. Some sections of this work are themselves original and not previously published before the *QRB* paper. These include two new sections to the appendix (Sections A.5 and A.6), the simplified derivation of the golden rule and its extension to exponential time course (Sections A.1.3 and A.7) and the suggested alternative treatment of the Beitz & Miller data in Section 5.6.4.

Numerous errors found in the *QRB* paper have been corrected.

The help of some individuals has been acknowledged in the text at appropriate places. I owe thanks for the initial encouragement to Britton Chance and to many associates for continued encouragement. Prepublication manuscripts were shared with me by many, including: Robert Blankenship, Dmitrii Chernavskii and Nina Chernavskaya, William Cramer, George Feher, Hans Frauenfelder and his group, Vitalii Goldanskii, Harry Gray, I. C. Gunsalus, John Hopfield, Shigeru Itoh, Toshiaki & Hiroko Kakitani, Bacon Ke, Rudolph A. Marcus, Noboru Mataga, Paul Mathis, T. Matsuo, John Miller, Mitsuo Nishimura, Tomoko Ohnishi, William Parson, F. G. Petrov, Andrei Rubin and his group, Akinori Sarai, Norman Sutin, Jens Ulstrup and Sigeo Yamosa.

I had useful conversations and/or correspondence with Lloyd Barr, Britton Chance, Peter Debrunner, Govindjee, Vitalii Goldanskii, Robin Hochstrasser, Barry Honig, Joshua Jortner, Toshiaki Kakitani, Kenneth Kaufmann, Rudolph A. Marcus, John Miller, Akinori Sarai whose association with me over the past seven months has been especially helpful, Norman Sutin and Robert Uphaus among others.

I am grateful to the staff of Cambridge University Press for their help and advice.

My wife deserves credit for her patient tolerance while I neglected household tasks during the several years I studied for and worked on this book.

This work was supported by the National Science Foundation, grants PCM78-23194, and PCM81-09854.

D. DeVault
1983

1

Introduction

'Tunnelling' is the metaphorical name given to the process, possible in quantum mechanics, but not in classical mechanics, whereby a particle can disappear from one side of a potential-energy barrier and appear on the other side without having enough kinetic energy to mount the barrier. One can think of this as a manifestation of the wave-nature of particles. The wavelength is larger if a particle is lighter. In particular, electrons, being very light compared to atoms, have wavelengths as large or larger than atoms at energies found in the valence shells of molecules. Thus, they easily ooze through and around atoms and molecules. We are also concerned with the tunnelling of heavy particles: nuclei, atoms, molecules.

It is not always a single particle and a well-defined barrier that is involved. The 'barrier' may be simply an arrangement of a whole set of atoms and molecules the potential energy of which is greater than the total energy available in the system. As long as a system cannot get from its initial state to its final state without passing through some classically forbidden arrangement we will call it 'tunnelling'.

The transfer of electrons from one molecule to another is a very important process in biology. Although early biochemists tended to think of oxidation-reduction reactions in terms of exchange of H or O atoms, it is clear that cytochromes and iron–sulphur centres, among other possibilities including certain chlorophylls, pheophytins and quinones, receive and transmit electrons directly. In both respiration and photosynthesis the primary action of the energy source (combustion of substrate by oxygen in respiration and absorption of light by chlorophyll or bacteriochlorophyll in photosynthesis) is to move electrons in an electron-transport chain consisting of such electron-transfer molecules. The biological system then extracts energy from this electron transport in order to phosphorylate ADP to ATP, or to move ions across membranes. ATP stands for adenosine triphosphate, a nucleic acid combined with pyrophosphate. Most cell processes obtain their energy by hydrolysing one of the pyrophosphate bonds to give adenosine diphosphate, ADP, and inorganic phosphate. Phosphorylating ADP to ATP replenishes the supply of immediately available energy.

Early theories of the mechanism of biological electron transport included semiconduction (Szent-Györgyi, 1941; Pullman & Pullman, 1963; Taylor, 1959; Rosenberg, 1965; Cope, 1965*a*, *b*) and attempts to find pathways through chains of conjugated and aromatic molecules (Winfield, 1965). As late as 1973 Dickerson and co-workers (Takano *et al.*, 1973) proposed a pathway for the electron through the protein part of cytochrome *c* involving hopping from one aromatic side-chain to another. See also Butler *et al.* (1975). A recent study of semiconduction effects induced in a protein is reported by Pethig & Szent-Györgyi (1977). Chance & Williams (1956) proposed rotation of cytochromes between electron transfers so that the haem could be close to donor when receiving the electron and close to acceptor on transmitting.

New insights into the mechanism of biological electron transfer were opened by the discovery that it can sometimes occur at temperatures too low for ordinary chemical reactions. Calvin & Sogo (1957) observed the production of unpaired electrons in photosynthetic systems exposed to light at 133 K. Witt *et al.* (1960*a*, *b*) observed light-induced spectral transients at 113 K in green plants and algae. Although 'electron exchange' was suggested as a possibility among others, this seems unlikely in view of their evidence of origin in triplet states and present knowledge of the fate of triplet states in photosynthesis (Mathis, 1969; Witt & Wolff, 1970; Wolff, 1975; Monger, Cogdell & Parson, 1976; Kung & DeVault, 1976). Chance & Nishimura (1960) clearly observed light-induced cytochrome oxidation in the photosynthetic bacterium *Chromatium vinosum* at liquid nitrogen temperature (77 K). Shortly afterwards Arnold & Clayton (1960) observed light-induced chlorophyll oxidation in dried films of chromatophores from the photosynthetic bacterium *Rhodopseudomonas sphaeroides* at liquid-helium temperature (4 K). Witt, Müller & Rumberg (1961) (Müller & Witt, 1961) observed cytochrome photo-oxidation in spinach chloroplasts at 123 K. Chance & Bonner (1963) observed light-induced cytochrome oxidation in green leaves at 77 K.

The understanding that electron transfer in biological systems involves quantum-mechanical tunnelling, as already conceived in physics and chemistry, came with the discovery that the rate of cytochrome oxidation in *Chromatium* is absolutely temperature-independent ($t_{\frac{1}{2}} = 2.3$ ms) from

Fig. 1.1. Temperature dependence of photosynthetic electron transfers in bacteria. Curves A and B: photo-induced cytochrome oxidation. Curves C and D: reversed primary reaction. (A) *Rhodopseudomonas* sp. *N.W.*, data of Kihara & McCray (1973) taken from their Fig. 6. (B) *Chromatium vinosum* (strain D), data of DeVault & Chance (1966) and DeVault *et al.* (1967), whole cells. ▲, data of Dutton *et al.* (1971), subchromatophore preparation, taken from their Fig 3; ×, data of Hales (1976) taken from his Fig. 1. (C) *Rhodospirillum rebrum.* ---, data of Parson (1967), taken from his Fig. 15; ●, data of Loach *et al.* (1975) taken from their Fig. 7, chromatophore preparation; ▲, same source, subchromatophore preparation. (D) *Rhodopseudomonas sphaeroides.* ⊙, data of McElroy *et al.* (1974) taken from their Fig. 3, reaction-centre preparation; □, data of Clayton & Yau (1972), taken from their Fig. 10, reaction-centre preparation; △, data of Hsi & Bolton (1974), taken from their Fig. 3, chromatophores and reaction centres, mostly treated with ortho-phenanthroline. ○, data of Romijn & Amesz (1976), from their Fig. 3; reaction-centre preparation. The horizontal scale to the left of $1/T = 0.01$ ($T = 100$ K) is strictly reciprocal temperature, i.e. Arrhenius plot. To the right of $1/T = 0.01$ the scale is linear in T. Also: d (scale)/dT is continuous at $1/T = 0.01$.

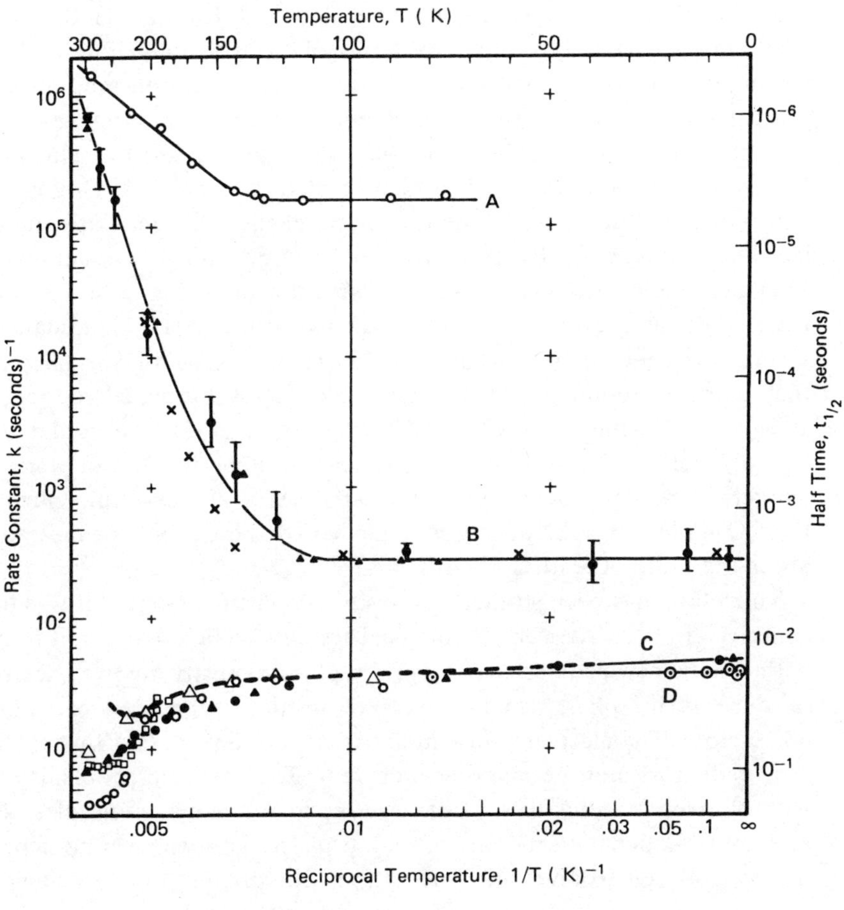

100 to 4 K (DeVault & Chance, 1966; DeVault, Parkes & Chance, 1967). See Fig. 1.1. Similar temperature independence of the rate of return of electron from acceptor to bacteriochlorophyll in *R. sphaeroides* was observed to 1.7 K by McElroy, Feher & Mauzerall (1969). See Fig. 1.1. The association of tunnelling with temperature independence does not infer that tunnelling does not occur when the rate is temperature-dependent. It is just easier to recognize as tunnelling when both slow and temperature-independent. The argument is that the slowness of the electron transfer (2 ms for *Chromatium* cytochrome oxidation below 100 K) indicates some kind of barrier to the transfer. The zero activation energy indicates that the process must be tunnelling the barrier. The interest of theorists was aroused and five notable attempts to explain the data followed at 2-year intervals: Grigorov & Chernavskii (1972), Hopfield (1974), Jortner (1976), Kuznetsov, Søndergard & Ulstrup (1978), and the last is really two contributions in different directions from the same laboratory: Yamosa (1980) and Kakitani & Kakitani (1981).

While we originally thought in terms of electron tunnelling a closer examination of the theory will show that nuclear motion must usually accompany it and the temperature independence at low temperatures may actually show that the nuclei (i.e. whole atoms) are tunnelling the barrier to their motion. Such was suggested by Levich & Dogonadze (1961) and by Sutin (1962). The tunnelling of the electron is ascertained by other arguments. Evidence for nuclear tunnelling in reactions of biological molecules or biological systems not involving either electron transfer or simple proton transfer has recently appeared. Goldanskii, Frank-Kamenetskii & Barkalov (1973) (Goldanskii, 1979*a*) observed the temperature-independent polymerization of formaldehyde below 10 K and Frauenfelder (1979) (Alberding *et al.*, 1976) have found that ligands such as CO bind to haem in several steps the last of which is temperature-independent below 10 K and clearly involves tunnelling of the CO molecules. The distances tunnelled by these atoms or molecules are of the order of tenths of angstroms.

Tunnelling has been studied in physics and chemistry since it was first conceived in 1927. As seen in biological systems, so far, however, it brings with it some new problems. The work in physics mostly involves electron movements in *solids*, usually layers of insulators between crystalline conductors. The electrons flow in a stream, are delocalized throughout the conductors and measured as a current. The chemical problems deal with electron transfer to or from *centres* in individual molecules, and attempt to understand the rate of such transfer. This was seen to depend mostly on environmental fluctuations involving low frequencies

($1-100\,cm^{-1}$) required to bring about an activated state necessary for the electron transfer and very little on the electron itself. The biological systems seen so far involve characteristics from both fields and are different from both. The electron transfers in the biological systems are from one molecular centre to another as in the chemical case, but they do not depend so strongly on environmental fluctuations since the centres are embedded in a more-or-less solid matrix – usually a lipid-protein membrane. Instead one sees a prominent role for high-frequency (e.g. $400\,cm^{-1}$) intramolecular vibrations. It appears that the biological systems are 'engineered' with respect to their vibronic couplings to produce special advantageous effects. Furthermore, there is considerable interest in the behaviour and path of the electron in the biological systems as in the physical systems. Path lengths of one or more nanometres have been postulated. Unlike the physical systems one sees transient behaviour as the electrons jump between specific centres instead of a 'tunnelling current'.

The experimentalists in the biological field have shown a need for theoretical assistance from experts on tunnelling in the physical and chemical fields and the special problems of the biological systems have challenged these experts. To fill this need for interdisciplinary communication, a conference on tunnelling in biological systems was held in Philadelphia, 3–5 November 1977, at the Johnson Foundation of the University of Pennsylvania. This book attempts to summarize the present problems in the field as revealed by the discussions at this conference and more recent work, and will draw upon them for inspiration. It is not, however, a report of the conference. References to the book reporting the conference (Chance *et al.*, editors, 1979*a*) will be designated '**T**' in the list of references to save space. An attempt will also be made here to explain in simpler terms some of the concepts which were shown to be important and to need explanation.

Other works discussing the application of the theories of Marcus (1965), Hopfield (1974) and Jortner (1976) to photosynthetic systems are papers by Marcus (1979*b*), Blankenship & Parson (1979), Okamura, Isaacson & Feher (1979*b*) Okamura *et al.* (1979*a*) and Jortner (1980).

2

Tunnelling in physics

2.1 Historical

Within three years of the invention of the modern quantum mechanics in 1925 the idea of tunnelling had been formulated and applied to: (1) intramolecular rearrangement (NH_3 inversion, etc.) by Hund (1927), (2) the field emission of electrons from a gaseous atom by Oppenheimer (1928), (3) the thermal emission of electrons from hot filaments by Nordheim (1928), (4) the field emission of electrons from metals by Fowler & Nordheim (1928), (5) alpha-particle radioactivity by Gurney & Condon (1928) and by Gamow (1928), and (6) the hydrogen molecule-ion by Pauling (1928). Within another year Bourgin (1929) proposed a tunnelling mechanism for unimolecular reactions such as the decomposition of N_2O_5. None of these used the word 'tunnelling', but they all, with the possible exception of Pauling, concerned themselves with the problem of a particle or particles, penetrating a potential energy barrier. See Fig. 2.1 which comes from Gamow's work.

The probability of the particle penetrating the barrier of Fig. 2.1, if small compared to 1, is:

$$P = 16EV(E + V)^{-2} \exp[-2b(2mV)^{\frac{1}{2}}/\hbar], \tag{2.1}$$

where $\hbar$ is Planck's constant divided by 2π and m is the mass of the particle. Another way of expressing this probability, used by Oppenheimer (1928), for example, is in terms of an element, H_{ab}, of the quantum-mechanical Hamiltonian matrix which governs the rate of transition from an initial state, a, of the system to a final state, b. The rate is proportional to the square of the absolute value of the matrix element: $|H_{ab}|^2$. It is often expressed by

$$\text{rate}(a \rightarrow b) = \frac{2\pi}{\hbar}|H_{ab}|^2\rho, \tag{2.2}$$

which has been called the 'golden rule' of quantum mechanics. ρ is a density of states, that is, the number of substates qualifying as b per unit interval of energy. In equation (2.2) the 'transition' may or may not be by tunnelling.

The hydrogen molecule-ion, H_2^+, was calculated earlier by Burrau (1927) but he did not carry his calculations out to nuclear separations at which tunnelling could be said to occur. Pauling also was more concerned with the properties of the molecule formed and with the importance of the exchange integral than with the fact that at large nuclear separations the electron has to pass through a potential-energy barrier. However, his equations are quite suitable for studying the latter problem too (see Section 5.6). The earlier calculation of the resonance energy of the electron-pair bond by Heitler & London (1927) might also involve tunnelling at larger than normal nuclear separations, but it is more difficult to see this because of the complications of having two electrons.

Pauling (1930) mentions the possibility of penetration of barriers to rotation by molecules rotating in crystals at temperatures too low for thermal activation over the barriers.

Gurney (1931) applied the concept of tunnelling to electrolysis. He made use of what is known as the Franck–Condon principle. Franck (1925) and Condon (1926), on the basis of the old quantum theory, had suggested that when an electron jumps from one quantum level to another in a molecule of more than one atom, it does so in a time too short for interatomic distances or velocities to change. Thus, even if nuclear separations were at equilibrium in the initial state, this may not correspond to equilibrium for the final state but rather to vibrational excitation. Condon (1928) updated the idea to conform with the new quantum theory,

Fig. 2.1. Quantum-mechanical tunnel penetration of a barrier. A plot of potential energy v. distance for a symmetrical rectangular barrier. E is the kinetic energy of the approaching particle; V is the barrier height above the particle energy; b is the barrier width. (After Gamow, 1928.)

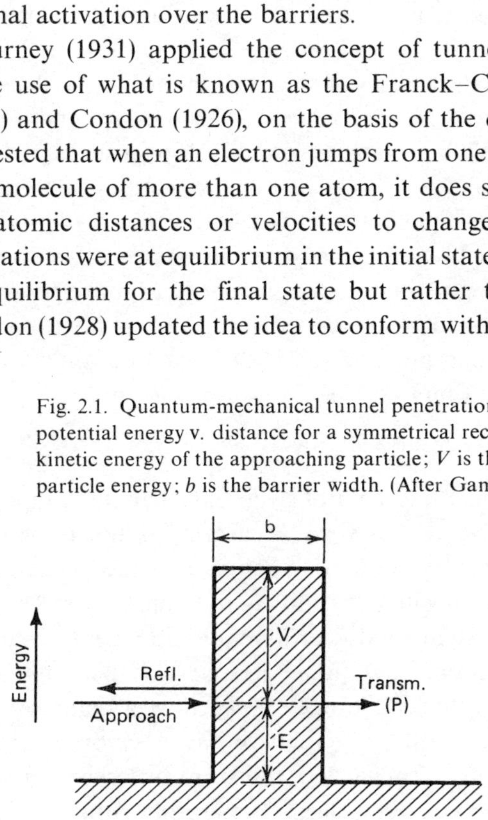

suggesting that the probability of the electronic transition is proportional to the square of the overlap integral of the vibrational wave-functions for initial and final states.

Gurney's idea was that when a charged, tunnelling particle, such as an electron, is present at a locality, the equilibrium polarization of the surrounding medium is different from when the particle is absent. As in the Franck–Condon principle, however, the part of the polarization due to positions of heavy particles does not have time to change during the moment that the electron tunnels from one position to another. Thus the immediate initial and final states of the tunnelling act are not likely both to have, equilibrium polarization. The non-equilibrium polarization, or 'reorganization', energy involved here runs like a theme through all subsequent electron-transport theory in condensed media. We will discuss this further in Chapter 3.

Nordheim (1932), Frenkel & Joffé (1932) and Wilson (1932) each applied tunnelling to the study of rectifying barriers. Zener (1934) used it to explain avalanche breakdown in semiconductor junctions. However, later developments showed that these efforts were a little premature and not quite correct.

Consideration of the polarization of the medium around a charge led Pekar (1951, 1954) to the concept of a 'polaron'. This was further developed by Holstein (1959*a*, *b*; 1979), among others. An electron in a particular location in a liquid or solid will polarize the surrounding molecules in such a way that the induced field is attractive to the electron and stabilizes its position. It is as though the electron had dug its own well and sits in it, 'self-trapped'. It does not have enough energy to move to another location unless the polarization moves with it. The combination of an electron and its surrounding polarization is called a 'polaron'. A review of the subject is given by Emin (1975).

2.2 The tunnel-diode

Next we turn to measurement of electrical currents, I, tunnelling through thin insulating layers. They are measured as functions of the voltage, V, applied to the conductors on each side of the insulator and at temperatures from near absolute zero to room temperature. There are two types of structure: p–n semiconductor junctions and metal-insulator–metal sandwiches in which one or both metals may be super-conducting. In 1957 Esaki (1958) invented the tunnel-diode, now a common semiconductor device in electronics. Its I–V curve has a region of negative resistance as shown in Fig. 2.2. Giaever (1960*a*, *b*) observed tunnelling electrical currents through oxide insulators separating superconducting metals. Mead and co-workers (Mead, 1969; Kurtin, McGill & Mead, 1971) studied electron tunnelling through barriers nearly 100 Å thick and confirmed the quantum-mechanical relationships as have also Mann & Kuhn (1971), Kuhn, Möbius & Bücher (1972). Fig. 2.1 would apply to these systems.

Fig. 2.2. Tunnel-diode (A) Energy-level diagrams along an axis across a *p–n* semiconduction junction. The applied voltage across the junction is 0, as shown by the equality of Fermi levels on either side. (B) Current *I* as a function of applied voltage *V*. Positive *V* corresponds to raising *n*-type energy levels relative to *p*-type in the diagram of part A.——, Tunnel-diode characteristic. The steep slope at $V = 0$ is due to the tunnelling. The fall off at $V = a$ results from the bottom of the conduction band on the right side going above the top of the valence band on the left. All conduction electrons in the *n*-type and all holes in the valence band of the *p*-type are then opposite band-gap across the junction. At $V = b$ the band boundaries are practically straight across and conduction does not require band jumping.---, Represents ordinary *p–n* rectifier. The barrier width is too large for tunnelling.○○○○○, A 'backwards diode' (actually a poor tunnel-diode).

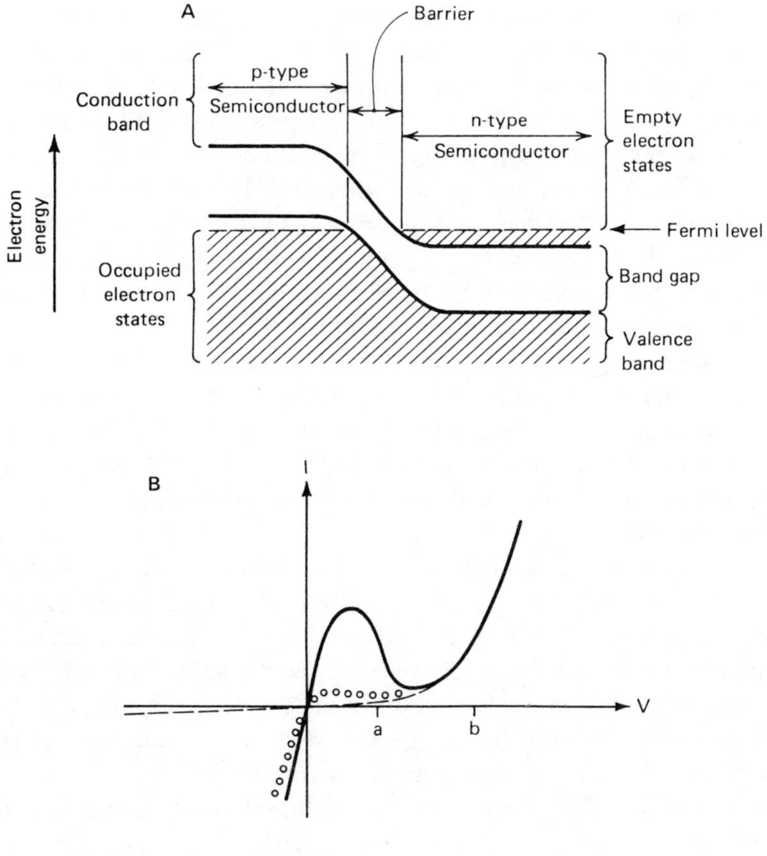

2.3 Phonon interactions

Holonyak *et al.* (1959) and Esaki & Miyahara (1960) first observed breaks in the current – voltage curves at low temperatures. These were interpreted as the onset of phonon-emission processes (i.e. excitation of vibrations) during the electron tunnelling. This inelastic process enhances the tunnelling current by allowing a larger range of initial and final energies by which the electron can satisfy overall conservation of energy.

Giaever (1960*a*, *b*), Giaever *et al.* (1962), Nicol, Shapiro & Smith (1960), Shapiro *et al.* (1962), Douglass (1961), Rowell, Chynoweth & Phillips (1962) and Rowell, Anderson & Thomas (1963) observed similar break-points in their observations of tunnelling to superconductors. These were interpreted as observation of the band gap predicted for superconductors by the Bardeen–Cooper–Schrieffer (1957) theory of superconductivity and also of the density of phonon states in the metals (Schrieffer, Scalapino & Wilkins, 1963). The previous one-electron theories of tunnelling had to be extended to include many bodies in order to explain the results of these experiments (Bardeen, 1961; Cohen, Falicov & Phillips, 1962).

Interaction of the tunnelling electrons with vibrations of molecules in the barrier were observed by Jaklevic & Lambe (1966). The metal-oxide insulating layer (between two metals) was deliberately doped with minute amounts of substances whose vibrational modes were to be observed. The break-points corresponding to excitation for specific vibrations appear as peaks if the second derivative, d^2I/dV^2, of the $I-V$ curve is plotted. This method has been developed by Coleman and associates (Coleman, 1979; Hansma & Coleman, 1974; Korman & Coleman, 1977), and also by Klein *et al.* (1973), into a new technique which has been called 'inelastic electron-tunnelling spectroscopy' (IETS).

The physics of electron tunnelling through conductor–insulator–conductor sandwiches appear at first sight to represent the original idea of a free particle tunnelling a fixed barrier (Fig. 2.1). Of interest to the chemical and biological cases, however, is the fact that eventually it became necessary to take account of atomic (nuclear) motions also. See Scalapino & Marcus (1967).

Josephson (1962) predicted and it was confirmed (Anderson & Rowell, 1963; Rowell, 1963; Fiske, 1964) that electrons tunnelling through junctions in superconducting rings should exhibit certain quantum-mechanical oscillatory phenomena. This discovery is the basis for numerous devices (Clarke, 1974) for accurately measuring small voltages or magnetic fields such as the 'SQUID', for detecting infrared radiation, and for fast computer memories (Fulton, Dynes & Anderson, 1973).

In 1973, Esaki (1974), Giaever (1974) and Josephson (1974) were awarded

Nobel prizes for their contributions to the field. For further reading on tunnelling in the field of physics there are books by Duke (1969) and by Burstein & Lundqvist (1969) and papers by Schrieffer (1979*a*), Dynes (1979), Duke (1979), Coleman (1979), Budgor (1979) and Knox (1979) in Chance *et al.* (eds.) (1979*a*).

3
Tunnelling in chemistry

Hund's (1927) investigation of intramolecular rearrangement via tunnelling of the classical barriers, Bourgin's (1929) tunnelling mechanism for unimolecular reactions and Gurney's (1931) application of electron tunnelling to electrolysis have already been mentioned.

3.1 Electrochemical theory

Gurney attempted to explain 'overvoltage' (the extra electrode voltage above the equilibrium value necessary to get appreciable electrolytic current) as that needed to supply the energy of the non-equilibrium polarization of the surroundings which will be found immediately after electron transfer because the atoms will not yet have had time to rearrange to new equilibrium positions (Franck–Condon principle). Gurney's theory was 'non-adiabatic' in the quantum-mechanical sense of the term, meaning that the electron transfer was considered to be a quantum jump from one electronic state to another. The rate of the electrode reaction was taken as proportional to the probability of the electron penetrating a barrier (between metal and ion) as in Fig. 2.1.

Horiuti & Polanyi (1935), following Frumkin's (1933) criticism of Gurney's theory, improved the formulation by taking account of the adsorption of hydrogen atoms (in the hydrogen electrode) at the metal surface and also of the concomitant reverse current flow. They used a diagram like Fig. 3.1 to discuss the requirements for nuclear motion.

In Fig. 3.1 the potential energy for nuclear motion for the whole system of donor, acceptor and surroundings is plotted against 'nuclear configuration'. The latter should have as many dimensions as there are degrees of freedom for nuclear motion in the system but on paper the one available abscissa will have to serve as a representative axis of nuclear motion from initial equilibrium configuration, *A*, to final equilibrium configuration, *B*. Fig. 5.4 is a similar drawing simulating two nuclear coordinates by perspective. Fig. 3.1 A shows two such potential-energy curves. Curve *R* gives the potential energy of the system when the electron is in the donor. Curve *P* is for when the electron is in the acceptor. They are parabolas if the nuclear vibrations about the equilibrium position are harmonic, following Hook's law forces. The displacement from *A* to *B* represents the change in the equilibrium polarization of the surroundings, that is, the shifts of nuclear position caused by the electron moving from donor to acceptor.

The electron transfer is a jumping from *R* to *P*. The Franck–Condon principle requires the nuclear configuration to be the same immediately

Fig. 3.1. The role of nuclear motion in electron transfer. The ordinates represent potential energy of the nuclei of the whole system: donor + acceptor + surroundings. The curves should be hypersurfaces in a space of as many dimensions as nuclear coordinates (+ 1 for energy). The abscissa represents a combination of the positions of all the nuclei disturbed by the electronic transition and runs along an axis from the initial equilibrium configuration, labelled *A*, to the final equilibrium configuration labelled *B*. Curve *R* is the potential energy of the nuclei when the electron donor is reduced and the acceptor oxidized. *P* is the nuclear potential energy when the donor is oxidized and the acceptor reduced. Passage from *R* to *P* represents electron transfer and occurs in the vicinity of the crossover, nuclear configuration *C*. In the diabatic case at the left, passage through *C* does not usually cause transition from *R* to *P*. In the adiabatic case at the right it usually does, and interaction between *R* and *P* is so strong that appreciable splitting into curves 1 and 2 is seen at *C*; *R* rides over to *P* on the surface, 1. See text for further explanation and the section on charge-transfer for explanation of the arrows labelled CT.

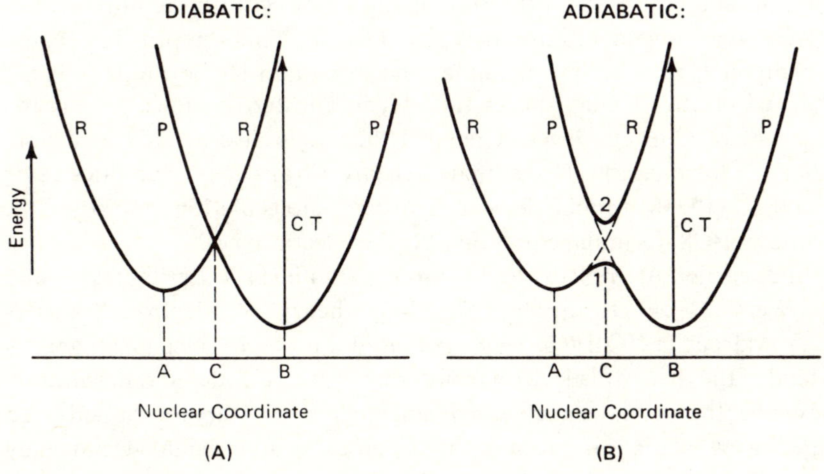

after the electron transfer as before. That is, the electron transfer itself must be a vertical transition in the diagrams of Fig. 3.1. Conservation of energy requires it to be a horizontal transition. The only possible way to meet both requirements is at nuclear configuration *C* where the two curves cross. The path of reaction in Fig. 3.1 A is then: activation from *A* to the crossing point at *C*. Then, depending on the probability of electron transfer when the nuclear configuration is at *C*, the system may jump to curve *P* and relax to *B*, or it may slide back down to *A*.

If the electron transfer at *C* is very probable the electron may go back and forth many times while the system is at *C*. This is the quantum-mechanical resonance of two electronic configurations having the same energy and it results in a splitting of the energy levels, one higher and one lower as shown in Fig. 3.1 B. In this case there is a smooth transition from *R* to *P* on a single curve as the nuclei go from *A* to *B* through *C*. This is called 'adiabatic', meaning that no quantum jump occurs. Horiuti & Polanyi's theory was adiabatic.

A clear explanation and organization of electrochemical theory at this stage of development is given by Randles (1952). Further development of electrochemical theory closely parallels that of the theory of electron transfer between ions in solution and I will not further discuss it separately.

3.2 Redox reactions in solution

Mott (1939, 1940) suggested that the oxidation of aluminium is controlled by electrons tunnelling through the oxide layer from the metal to absorbed O_2. This suggested to Libby (1940, 1979) that electron tunnelling could explain the rapid exchange of MnO_4^{-} and MnO_4^{2-} in solution and (1952, 1979) that tunnelling along with the requirement of solvent rearrangement to satisfy the Franck–Condon principle could explain such facts as, for example, that ferri- and ferrocyanide ions exchanged electrons much faster than ferric and ferrous ions. Apparently unaware of Gurney's electrochemical theory he also arrived at a non-adiabatic formulation. He estimated electron-transfer probabilities using Pauling's (1928) formulas for the hydrogen molecule ion instead of the Gamow (1928) formula, equation (2.1). See Section 5.6.2.

After earlier attempts by R. J. Marcus, Zwolinski & Eyring (1954) and by Weiss (1954) to apply tunnelling theory to electron transfers R. A. Marcus (1956, 1965, 1968, 1979*a*, *b*), in a series of papers, greatly extended the theory of electron transfer both at electrodes and in solution. He concentrated on the nuclear rearrangements necessary for the activated state, using statistical mechanical arguments, and included not only hydration shells of the ions but also the polarization effects in the solvent. Roughly, his results may be described as follows.

3.2.1 *Marcus' theory*

The rate constant, k in absolute reaction-rate theory (Eyring, 1935*a*, *b*) form is:

$$k = \kappa B \exp(-\Delta G^{\ddagger}/k_{B}T) = \kappa B \exp(\Delta S^{\ddagger}/k_{B}) \exp(-\Delta H^{\ddagger}/k_{B}T). \tag{3.1}$$

This relates to the Arrhenius equation:

$$k = k_{\infty} \exp(-E^{\ddagger}/k_{B}T) \tag{3.2}$$

if $E^{\ddagger}$ is identified† with $\Delta H^{\ddagger}$ and k_{∞} with $\kappa B \exp(\Delta S^{\ddagger}/k_{B})$. In Marcus' theory, κ, which involves the electron-transfer 'matrix element' as found in equation (2.2), takes care of the electron-transfer probability. It has the value 1 if the reaction is adiabatic and less than 1 if non-adiabatic. If the reaction is bimolecular, B is the liquid phase collision frequency, Z, about $10^{11}\ \mathrm{M}^{-1}\ \mathrm{s}^{-1}$. If monomolecular (intramolecular), B is a vibrational frequency, $k_{B}T/h$, about $10^{13}\ \mathrm{s}^{-1}$. The other symbols have their usual meanings: $\Delta G^{\ddagger}, \Delta S^{\ddagger}, \Delta H^{\ddagger}$ and $E^{\ddagger}$ are the free energy, entropy, enthalpy, and energy of activation respectively; k_{B} is the Boltzmann constant, $8.617 \times 10^{-5}\ \mathrm{eVK}^{-1}$ (if the energy quantities are per molecule, but should be interpreted as gas constant if the energies are per mole); T is the absolute temperature and h is Planck's constant. Marcus gives for $\Delta G^{\ddagger}$

$$\Delta G^{\ddagger} = (\lambda + \Delta G^{\circ})^{2}/4\lambda \tag{3.3}$$

for monomolecular reactions and adds a 'work' term, w, for any energy not directly coupled to the electron transfer but needed to prepare the reactants, such as bringing them together in the bimolecular case or possibly a conformational change in a monomolecular case. Biological electron transfers between protein molecules fixed in membranes correspond more closely to the monomolecular case and w is usually ignored. However, Schneider *et al.* (1980) have shown that proteins do diffuse laterally in membranes and so it is possible that w should not be ignored. ΔG° is the standard free-energy increase for the electron-transfer reaction.

The quantity λ is the reorganizational energy which is coupled to the electron transfer. It is the energy that would be required to move all the atoms from their equilibrium positions before electron transfer to the equilibrium positions they would have after electron transfer without,

† Strictly, $E^{\ddagger}$ should be $\Delta H^{\ddagger} + \frac{1}{2}k_{B}T$ for bimolecular or $\Delta H^{\ddagger} + k_{B}T$ for monomolecular reactions because of the temperature dependence of B, but the difference is ordinarily negligible. $k_{B}T$ is 0.03 eV at room temperature.

however, transferring the electron. (In the related process of energy transfer it is half the energy loss in the 'Stokes shift' between peak absorption and peak emission.)

Marcus (1965) separates λ into two parts, $\lambda_i + \lambda_o$, where λ_i is the reorganizational energy of the inner shell of atoms and λ_o is that of the surrounding solvent molecules. λ_i is calculated from the parameters of the inner-shell vibrational modes:

$$\lambda_i = \tfrac{1}{2}\sum_j k_{Hj} Q_j^2, \tag{3.4a}$$

if the modes are normal modes, or

$$\lambda_i = \tfrac{1}{2}\sum_{jk} k_{Hjk} Q_j Q_k \tag{3.4b}$$

for more general modes with, in either case, the Hook's law force constant, k_H, approximated by an average between its value before electron transfer, f_r, and its value after electron transfer f_p:

$$k_H = 2f_r f_p/(f_r + f_p). \tag{3.4c}$$

An expression like (3.4*c*) holds for each value of the subscripts in (3.4*a*) or (3.4*b*). The Qs are the displacement of the equilibrium positions of the vibrational coordinates caused by the electron transfer. Harmonic oscillations are assumed. A simple derivation of equation (3.3) with (3.4*a*) will be given in the theoretical section.

λ_o is estimated from the polarizability of the solvent considered to be a continuous polar medium (see Section A.5, equation (A.104)):

$$\lambda_o = \frac{(\Delta e)^2}{4\pi\varepsilon_o}\left(\frac{1}{2r_1} + \frac{1}{2r_2} - \frac{1}{r_{12}}\right)\left(\frac{1}{D_{op}} - \frac{1}{D_s}\right), \tag{3.5}$$

where Δe is the charge transferred from donor to acceptor; r_1 and r_2 are the radii of the two reactants when in contact; r_{12} is $r_1 + r_2$; D_{op} is the square of the refractive index of the medium and D_s its static dielectric constant, and ε_o is the permittivity of space to give SI units. In general one can see that λ involves moving positively charged polar groups in the vicinity of the electron donor further from the donor and in the vicinity of the acceptor closer to the acceptor (and vice versa for negatively charged groups).

It is seen that the emphasis in the theory of electron-transfer reactions moved away from consideration of the electron tunnelling to consideration of the accompanying nuclear motions. Yet both parts remain: κ in equation (3.1) for the electron part and $\Delta G^{\ddagger}$ (equations (3.1) and (3.3)) for the nuclear part.

From equation (3.3) it can be seen that for a series of reactions having the same value of λ and different values of ΔG° the rate should be a maximum when $\Delta G^\circ = -\lambda$ and should decrease if the reaction becomes more exothermic ($\Delta G^\circ < -\lambda$). This is sometimes called the 'energy-gap law', $-\Delta G^\circ$ being the energy gap.

In the case of self-exchange reactions $\Delta G^\circ = 0$, and equation (3.3) gives:

$$\Delta G_{11}^{\ddagger} = \lambda_{11}/4. \tag{3.6}$$

Similarly for $\Delta G_{22}^{\ddagger}$ where subscripts 11 and 22 refer to self-exchange reactions and 12 will refer to the cross-reaction involving the same pair of reactants. $\Delta G_{11}^{\ddagger}$ and $\Delta G_{22}^{\ddagger}$ and, therefore, λ_{11} and λ_{22} can be evaluated by equation (3.1) from the rate of the self-exchange reactions and their temperature coefficient.

Equation (3.3) for the cross-reaction is also written:

$$\Delta G_{12}^{\ddagger} = \frac{\lambda_{12}}{4} + \frac{\Delta G_{12}^{\circ}}{2} + \frac{(\Delta G_{12}^{\circ})^2}{4\lambda_{12}}. \tag{3.7}$$

Marcus (1965) shows that λ_{12} should be closely equal to the arithmetic mean of λ_{11} and λ_{22}. Thus, using (3.6), one can write (3.7) as:

$$\Delta G_{12}^{\ddagger} = \frac{\Delta G_{11}^{\ddagger}}{2} + \frac{\Delta G_{22}^{\ddagger}}{2} + \frac{\Delta G_{12}^{\circ}}{2} + \frac{(\Delta G_{12}^{\circ})^2}{8(\Delta G_{11}^{\ddagger} + \Delta G_{22}^{\ddagger})}. \tag{3.8}$$

The exponentials of these after division by $-k_B T$ converts the $G^{\ddagger}$ to rate constants and the G_{12}° to an equilibrium constant, K_{12}, giving the following relationship (Marcus, 1965) between rates of two different electron-exchange reactions and the rate of the corresponding cross-reaction:

$$k_{12} = (k_{11}k_{22}K_{12}f)^{\frac{1}{2}}, \tag{3.9}$$

where

$$\ln f = (\ln K_{12})^2/4 \ln\{k_{11}k_{22}/(\kappa B)_{11}(\kappa B)_{22}\}. \tag{3.10}$$

Often, f can be assumed to be close to one. In converting the $\Delta G^{\ddagger}$ to rate constants we failed to include work terms, w, but they can usually be considered small or cancelling.

3.2.2 *Other theories*

A modification was proposed by Hush (1958, 1961). He suggested that adiabatic electron transfer is not a Franck–Condon type process with a sudden electron jump but rather a gradual movement of the electron probability from donor to acceptor in synchrony with the nuclear passage through the activated configuration (Fig. 3.1B) and that the polarization of surroundings follows the charge passage in equilibrium with it at all points. His conclusions, however, are not greatly different from those of Marcus (Hush, 1961).

Sutin (1979) suggested that for the case of self-exchange, equation (3.1), with w included, can be recast into the form of an equilibrium constant for the formation of an activated complex multiplied by the rate constant expression appropriate for the monomolecular case which one has after formation of the activated complex. Sutin thus replaces κB in equation (3.1) with its monomolecular value, $k_B T/h$, and he compensates for this by adding to $\Delta G^\ddagger$ a term, $\Delta G_{\text{trans}}{}^\ddagger$, equal (by onc formula) to

$$-k_B T \ln(hZ/k_B T).$$

Levich & Dogonadze (1959, 1961) advanced a completely quantum-mechanical formulation, including quantization of the nuclear motions. They drew upon theories that had been developed by Huang & Rhys (1950), Lax (1952) and Kubo (1952) for radiative and non-radiative electronic transitions in solids coupled to vibrational excitations in the system and surroundings. This theory will be discussed further in Chapter 5. Their results approach those of Marcus at high temperatures but show distinctly quantum-mechanical effects at low temperatures. Levich (1966, 1970) has written two detailed reviews of the whole subject. Further work on the theory is found in Dogonadze *et al.* (1965, 1972*a*, *b*, 1973*a*, *b*, 1979), Ulstrup (1973, 1977, 1979), Ulstrup & Jortner (1975), Van Duyne & Fischer (1974), Fischer & Van Duyne (1977), Christov (1975), Kestner, Logan & Jortner (1974), Sarai (1979), Yamosa (1978), Englman & Jortner (1970), Brailsford & Chang (1970), Kakitani & Kakitani (1981) and Brunschwig *et al.* (1980).

3.2.3 *Comparison with experiment*

Experimental work to test the theories runs along several lines. One is the collection of data on rates of electron exchange between ions in solution. Chou, Creutz & Sutin (1977) carefully survey the results of various combinations of four oxidant (acceptor) ions and four reductants. These include Eu^{2+}(aq), V^{2+}(aq) and complex ions of Ru and Co. They conclude that the Marcus theory is reasonably successful in predicting rate constants for cross-reactions from those for self-exchange, but that the observed values are usually smaller. Sutin (1979) compares the measured rates of four self-exchange reactions with the values obtained from the Marcus equation using independent estimates of the various parameters. $\Delta G^\ddagger$ agrees well. $\Delta H^\ddagger$ agrees well at zero ionic strength but shows opposite dependence on ionic strength. $\Delta S^\ddagger$ agrees well with the formula derived by Marcus using partition function arguments but is much more negative than the value predicted by another formula derived from diffusion (Eigen, 1954) and also from free-volume considerations (Fuoss, 1958).

3.2.4 *Quantum-mechanical estimation of* λ_i

Sutin (1979) also reports *ab initio* quantum-mechanical calculations by M. D. Newton of the parameters for $Fe^{2+} - Fe^{3+}$ self-exchange. These include placing a Fe^{3+} ion and a Fe^{2+} ion 6.9 Å apart, surrounding each with $6H_2O$ molecules at octahedral coordinates and calculating the orbitals for all the valence electrons of the iron atoms in the fields of their cores plus electrostatic interaction with the water dipoles. The electron-transfer probability ('matrix element') is estimated from the overlap of the orbitals from the two iron atoms.

When Fe^{2+} changes to Fe^{3+} the water molecules move closer to the iron. This nuclear motion is the inner-shell reorganization whose energy is Marcus' λ_i. Sutin (1962) suggested that this motion involved nuclear tunnelling and we will refer to it again under that heading. Newton used a formula of Nikitin (1974) which estimates electron-transfer rate as a function of the electronic matrix element (Landau, 1932; Zener, 1932), and of the nuclear tunnelling. The calculation estimates that the electron transfer is non-adiabatic at 6.9 Å separation between the iron atoms but better agreement with experiment may be obtained by moving the iron atoms closer to the point where the transfer may become adiabatic. Further work is reported in Brunschwig *et al.* (1980). Adiabatic seems to be the rule for most electron transfer between ions in solution.

3.2.5 *Non-tunnelling temperature independence*

Note that reaction rates can be temperature-independent for other reasons than tunnelling. A good criterion for temperature independence is $\Delta H^\ddagger = 0$. This is not exact because of small temperature dependence of the non-exponential factors and of $\Delta H^\ddagger$ itself, but such are secondary to the exponential dependence.

Since $\partial(G/T)/\partial(1/T) = H$, dividing equation (3.8) by T and taking the partial derivative with respect to $1/T$ gives:

$$\Delta H_{12}^\ddagger = \frac{1}{2}\left\{\Delta H_{11}^\ddagger + \Delta H_{22}^\ddagger + \Delta H_{12}^\circ + \frac{\Delta G_{12}^\circ \Delta H_{12}^\circ}{2(\Delta G_{11}^\ddagger + \Delta G_{22}^\ddagger)} - \frac{(\Delta G_{12}^\circ)^2(\Delta H_{11}^\ddagger + \Delta H_{22}^\ddagger)}{4(\Delta G_{11}^\ddagger + \Delta G_{22}^\ddagger)^2}\right\}. \tag{3.11}$$

It is suggested (Sutin, 1979) that $\Delta S_{11}^\ddagger$ and $\Delta S_{22}^\ddagger$ should be small or zero so that $\Delta H_{11}^{\ddagger} + \Delta H_{22}^{\ddagger} = \Delta G_{11}^{\ddagger} + \Delta G_{22}^{\ddagger} = \lambda_{12}/2$ by equation (3.6), etc. Substituting λ_{12}, therefore, and also $\Delta H_{12}^{\circ} = \Delta G_{12}^{\circ} + T\Delta S_{12}^{\circ}$ into equation (3.11) gives:

$$\Delta H_{12}^{\ddagger} = \frac{\lambda_{12}}{4} + \frac{\Delta G_{12}^{\circ}}{2} + \frac{T\Delta S_{12}^{\circ}}{2} + \frac{(\Delta G_{12}^{\circ})(\Delta H_{12}^{\circ} + T\Delta S_{12}^{\circ})}{4\lambda_{12}}$$
$$= (\lambda_{12} + \Delta G_{12}^{\circ})[\lambda_{12} + \Delta H_{12}^{\circ} + T\Delta S_{12}^{\circ}]/4\lambda_{12}. \quad (3.12)$$

From equation (3.12) it is clear that $\Delta H_{12}^{\ddagger}$ will equal zero if either (*a*) $\lambda_{12} = -\Delta G_{12}^{\circ}$, obvious also from equation (3.3), or (*b*) $\lambda_{12} = -\Delta H_{12}^{\circ} - T\Delta S_{12}^{\circ}$.

Sutin (1979) shows that the oxidation of aqueous Fe^{2+} by $Ru(bpy)_3{}^{3+}$ where 'bpy' stands for bipyridyl, comes close to case (*b*). For this reaction, in units of kcal mol^{-1}, $\lambda_{12} = 36$; $\Delta G_{12}^{\circ} = -14$; $\Delta H_{12}^{\circ} = -27$, and $T\Delta S_{12}^{\circ} = -13$ at 298 K. Substituted into equation (3.12) these give $\Delta H_{12}^{\ddagger} = -0.6$ with $(\lambda_{12} + \Delta G_{12}^{\circ}) = 22$ and $(\lambda_{12} + \Delta H_{12}^{\circ} + T\Delta S_{12}^{\circ}) = -4$. The reaction has a slightly negative temperature coefficient, in fact. In the above equations we have neglected the work terms, *w*, and so what we have designated with superscript ‡, are designated with * by Sutin (1979), who reserves ‡ for the measured or measurable values which include the work terms. These terms do not change the main picture and are expected to be even more negligible in biological membrane systems. The unusual feature of the $Fe^{2+} + Ru(bpy)_3{}^{3+}$ reaction is the large $-T\Delta S_{12}^{\circ}$. This results almost entirely from Fe^{3+} binding its inner-shell $6H_2O$ more tightly than Fe^{2+} and the resulting large increase in vibrational frequency, causing wider spacing of vibrational levels.

A further discussion of non-tunnelling temperature independence is found in Section 5.4, on the role of low-frequency vibrational modes coupled to electron transfer.

3.3 Intramolecular electron transfer

A second line of experimental investigation is the measurement of rates of electron transfer between redox groups held together by a bridging ligand. By use of different types of bridging groups with different distances between the redox centres (usually metal ions) one can hope to get a better understanding of the needs of the electron transfer itself apart from the complications of bringing ions together. Taube has studied electron transfer in chemical systems for many years (Taube, 1975) and reports on a series with bifunctional bridging groups (Taube, 1979; Isied & Taube, 1973; Fischer, Tom & Taube, 1976; Rieder & Taube, 1977). They measured the rate of electron transfer from Ru^{2+} to Co^{3+} in such compounds and charge transfer or intervalence band light absorption in the corresponding Ru^{2+}, Ru^{3+} compounds. The latter type of measurement will be discussed in the next section. However, the strength of the absorption is related to the electron-transfer matrix element. When, as was found, most of the reactions were adiabatic, the rates were insensitive to the electron-transfer probability because κ in equation (3.1) was close to 1 anyway, but the intervalence band absorption was still sensitive. One series of bridging ligands, for example, was

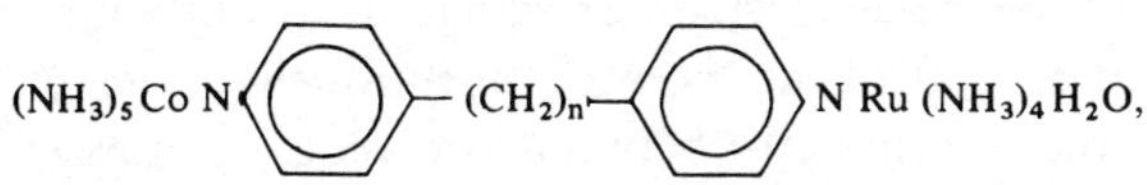

where n varied from 0 to 2. With $n = 1$ or 2 the electron transfer became non-adiabatic, but with $n = 2$ there is the possibility, and some evidence for it, that the molecule bends so that the two metals get close enough to transfer electrons directly instead of through the ligand. When the ligand is pyrazine (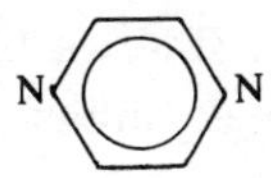) the rate becomes 100 times faster than expected for adiabatic. No explanation is given. See also data on a pyrazine-ruthenium mixed-valence complex by Drago and co-workers (Bunker *et al.*, 1978).

Isied & Kuehn (1979) investigated imidazolate anion () as a bridging ligand in the same types of measurement as in the preceding paragraph. They find the ligand to be effective in electron transfer and this should be interesting for biological studies.

Norton *et al.* (1979) studied the series:

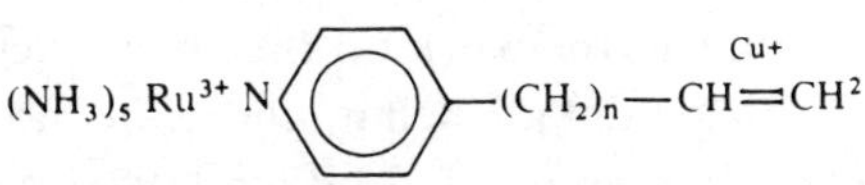

and also a similar series with the alkyl chain replaced by polyisoprenyl chains. The Cu^+ is loosely bound to the olefin bond and in equilibrium with free Cu^+ in solution. Direct measurement of rate of transfer by optical absorption changes and measurement of intervalence band strength can be done on the same molecule. With $n = 2 - 7$ or with the polyisoprenyl chains the rates are second order and very similar suggesting transfer to neighbouring molecules instead of intramolecular. With $n = 0$, first-order rates are observed at higher concentrations. They find no evidence that the olefin groups aid electron transfer.

Weissman (1958) and Voevodskii, Solodovnikov & Chibrikin (1959) used EPR to measure the exchange of electrons between two aromatic groups held apart on the same molecule by chains of CH_2 groups of various lengths. They found, in general, that an extra electron on such molecules (radical anions) resonated between the two aromatic groups at least 10^7 times/s if they were separated by no more than about two or three CH_2 groups.

Yamosa (1978) developed basic equations to describe the effect of polar solvents on intramolecular electron exchange in a solute molecule. The theory is related to that of the 'polaron' mentioned earlier. A molecule containing a plane of symmetry between two redox groups will show a resonance between the two groups if one is oxidized and the other reduced and the molecule is in a vacuum. In a polar solvent, however, the symmetry may be broken if the solvation energy is larger than the resonance energy. The electron can become trapped on one side or the other by unsymmetrical polarization of the solvent favouring the unsymmetrical localization of the electron. His calculations for the excited state of bianthryl were borne out by picosecond flash spectroscopy by Nakashima, Murakawa & Mataga (1976), who found that the spectrum of the excited state was like anthracene when the bianthryl was dissolved in hexane but like a superposition of anthracene anion and cation when in acetone.

Mataga and co-workers have used many techniques to make extensive studies of the kinetics of charge transfer and intramolecular excimer formation in pairs of organic donors and acceptors tied together by hydrocarbon chains (Hinatu *et al.*, 1978; Okada *et al.*, 1977; Migita, Kawai & Mataga, 1978; Hayashi *et al.*, 1977*a*, *b*; Mataga, Migita & Nishimura, 1978). They find that if the chains are loose enough (more then two CH_2 groups) the photo-excited molecules tend to fold over into a sandwich-type structure to produce an excimer. The effects of solvent polarity are studied. Some of the work on the heteroexcimers has been done in micelles in order to get data which might relate more closely to biological systems (Waka, Hamamoto & Mataga, 1978) and which might enable the heteroexcimers to be used as probes of membranous systems.

3.4 Charge-transfer (intervalence) bands

If a molecule or complex containing both donor and acceptor rests in its ground state with the electron in the acceptor, the state with the same nuclear configuration but with electron in donor could be reached by absorption from light of the difference in energy as shown by the arrows labelled CT in Fig. 3.1. The resulting absorption band was called a 'charge-transfer' band by Mulliken (1952) and an 'intervalence' band by Hush (1967; Allen & Hush, 1967), who restricted the second term to cases in which the electron can be said to localize definitely in donor or acceptor. (More general cases may involve only partial transfer.) The intensity of the band is related to the electron-transfer matrix element, H_{ab}, and thus is an independent source of information concerning its value. The theory will be discussed in more detail in the appendix. A review of the subject is given by Robin & Day (1967). Applications to electron transfer between metal complexes were cited in the last section.

3.5 Pulse radiolysis

Much information on electron transfer has come from pulse radiolysis studies. In a typical experiment a 25 mM solution of a substance able to accept electrons is frozen to a rigid glass in liquid nitrogen and subjected to a brief pulse of high-energy electrons from an accelerator. Many electrons (but fewer than the number of solute molecules) are trapped in the sample and can be assayed by their light absorption in the near infrared which gives them a light blue colour.

The colour fades as the trapped electrons tunnel to the scavenging molecules. The time course of the fading is measured. When a given fraction of the absorption has disappeared one knows that a certain fraction of the electrons have found acceptors. Assuming that this fraction includes all trapped electrons less than a certain distance from an acceptor Beitz & Miller (1979*a*, *b*) calculate what the distance is. They obtain an unambiguous tunnelling-time v. distance relationship. (See equation (5.108)). They used this technique with 48 different scavenger solutes dissolved in methyl tetrahydrofuran. The fastest was benzonitrile, which accepts electrons from 25 Å in 10^{-6} s and from 35 Å in 0.03 s. The slowest was benzoquinone which takes 1 s at 25 Å and 100 s at 30 Å. The dependence of rate on distance was found to agree with the Gamow exponential function, equation (2.1).

More recent studies (Miller, Beitz & Huddleston, 1982) suggest that some of the temperature dependence of electron transfer at low temperatures could result from a temperature-dependent slowness of repolarization of the surroundings following the initiating pulse. Thus in photosynthetic systems the nuclear configuration under which secondary electron transfer occurs may be only partially readjusted following the initial photo-induced transfer.

Van Leeuwen (1981) also presents extensive work with pulse radiolysis and, in his chapter 6, a detailed theory in which the Franck–Condon factor depends on the distance between the electron trap and the scavenger ion.

3.5.1 *Energy-gap law*

In a series of reactions in which all other parameters, including λ, remain constant while ΔG° decreases from zero, the rates should increase until $\Delta G^\circ = -\lambda$ and then should decrease according to equations (3.1) and (3.3). More exact quantum theories, for example Van Duyne & Fischer (1974), Ulstrup & Jortner (1975), predict similar behaviour except that the fall-off of rate after going over the maximum is more gradual than the rise before maximum because of increased absorption of the excess energy by vibrations.

However, Rehm & Weller (1970) reported a series of experiments showing no decrease in rate upon further decrease of ΔG° after the maximum. A number of other workers found a similar failure of equation (3.3) while more recent measurements (Van Duyne & Fischer, 1974; Creutz & Sutin, 1977; Miller, 1978) showed some decrease of rate after the maximum but not as much as predicted.

Beitz and Miller's data showed rather good adherence to equation (3.3), especially if allowance is made for many cases in which the product state is not the ground state but an excited electronic state whose $-\Delta G^\circ$ is smaller. An example of the decrease in rate due to energy gap is the data already mentioned for benzonitrile whose ΔG° matches $-\lambda$ at -1.1 eV and for benzoquinone whose ΔG° is estimated at -3.0 eV. The success of the pulse-radiolysis work in detecting adherence to the energy-gap law is attributed to the frozen matrix in which the reactions occur, which eliminates complications resulting from ion mobility. This solid-state set up more nearly approximates what one can expect in the biological systems which have been studied. It is interesting to see that such systems involve interaction with their vibrational modes through the coupling factor λ in the manner suggested by equation (3.3).

Other pulse radiolysis data in the chemical field include: Brocklehurst (1973), Brocklehurst, Bull & Evans (1975), Miller (1972, 1973, 1975), Miller & Willard (1972), Buxton & Kemsley (1976), Henglein (1974, 1975), Grätzel, Henglein & Janata (1975) and Frank *et al.* (1976). Reviews have been written by Brocklehurst (1979), Hart & Anbar (1970), Rice & Pilling (1978), Hunt (1976), Kevan (1974), and a conference on 'Electrons in Fluids' (Proceedings, 1977).

Marcus & Siders (1982) consider the inverted region ($-\Delta G^\circ > \lambda$) in some detail. They calculate diffusion-limited rates which tend to hide the inverted effects.

3.6 Miscellaneous chemical electron-transfer data

Some miscellaneous early chemical data pointing to electron tunnelling are the following. (i) Hornbeck (1952) and Holstein (1952) noticed that gaseous atoms and ions exchange electrons over distances almost twice as large as those required for momentum transfer in collisions. (ii) Eley *et al.* (1953) (Eley & Parfitt, 1955; Eley & Willis, 1961) suggested tunnelling between molecules to explain the electrical conductivity properties of organic crystals. Attempts had been made to treat organic crystals as semiconductors but there were difficulties with this picture since the compounds in question were more like insulators. Any electrical conductivity required high activation energy. It may be worthwhile to digress a little about semiconductivity since the concept has also been proposed for biological systems.

The concept of conductivity in semiconductors is that an electron in an excited state of one atom in a non-metallic crystal can move to the next atom and thus wander around the crystal. This possibility allows the construction of new wave-functions for such an electron from the wave-functions of each atom. The new combination wave-functions each extend the whole width of the crystal. They are spread, by the strength of the interaction between the original atomic orbitals, over a range of energies making a band, called the 'conduction band'. This band structure is shown in Fig. 2.2. The band has as many constituent wave-functions as there are atoms in the crystal so the spacing between levels in the band is practically infinitesimal and the electron has much freedom to manoeuvre (to change motion by jumping from one wave-function to another nearby). If, however, as in an insulator, the interaction between the atomic orbitals is weak, meaning that the electron has difficulty going from one atom to the next, the spread in the energies of the constituents of the conduction band will be small. The band will be narrow and the mobility of electrons in it will be small. The electron motion may then be better discussed in terms of a hopping from one atom or molecule to the next, rather than as motion in a band. The hopping could be by thermal activation to overcome the barriers between the molecules. Eley's suggestion was that the hopping could also be accomplished by tunnelling. A further development and more technical discussion of this question is given by Glaeser & Berry (1966). (iii) There is older data by Horne (1963) that ions in water or frozen in ice can exchange electrons when they average 100 Å apart. Ruff (1965) proposed electron tunnelling to explain it. (iv) Miller, Hartman & Abrash (1982) observed the quenching of phosphorescence by electron transfer over centre-to-centre distances of 19–30 Å in frozen organic glasses at 77–196 K.

Studies of photo-induced electron transfers and also electron exchanges have been made in micelles by Matsuo and co-workers. These involve various redox components. (Kano *et al.*, 1978; Takuma, Shuto & Matsuo, 1978; Takuma *et al.* 1981).

3.7 Nuclear tunnelling

The importance of 'nuclear' (or 'atomic' or 'molecular') tunnelling as a process which may accompany electron tunnelling and as a path for chemical reaction not involving electron transfer has been mentioned. The concept is basically the same as that pictured in Fig. 2.1. One simply sets m equal to the atomic or molecular mass in equation (2.1). The much larger m means that the distances, b, must be much smaller to give rates conforming to a human time scale (0.01 nm instead of 1 nm, order of magnitude with usual barrier heights, V). In Fig. 3.1 (A or B), nuclear tunnelling would mean a horizontal movement from R to P below the level at which the two curves cross. This appears at first to be a violation of the Franck–Condon principle requiring that sudden changes must be vertical. Fig. 3.2 shows how there is actually no violation. It shows wave-functions (squared) for the system vibrating in several different states with total energy less than the activation energy, $E^{\ddagger}$, necessary to go over the barrier at configuration C. It is seen that the quantum mechanics, in contrast to classical mechanics, still gives a small probability that the system can be found at C below the top of the barrier. At C, the electron can transfer, if such is contemplated, to a vibrational wave-function (not shown) extending from the P state. Thus with or without electron transfer the change from R to P can be said to occur at C, even below the top of the barrier. However, the true expression of the Franck–Condon principle is the overlap of the wave-functions from R and P (see equation (5.36)). In the case of tunnelling it is simply the exponential tails of the wave-functions that overlap. (See Section 5.6.1(a) for further discussion.)

Fig. 3.2. Nuclear tunnelling. Nuclear vibrational energy states are pictured in electronic state R. Their wave-functions reach across configuration C even when the total energy is less than the potential energy at C.

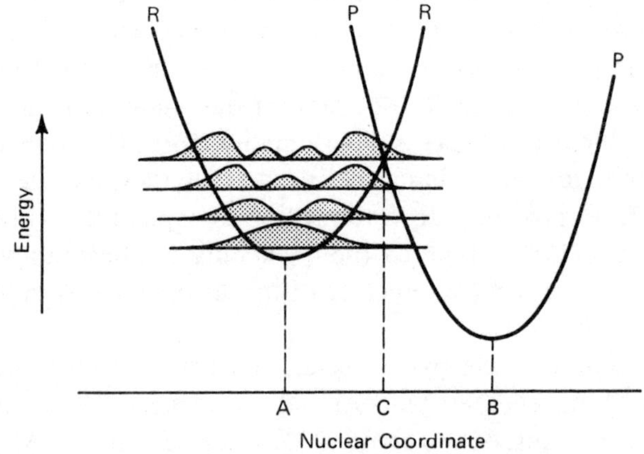

The rate of nuclear tunnelling has a temperature dependence consisting of three regions. At very low temperatures the molecules in the *R* state of Fig. 3.2 will all be in the lowest vibrational state. If this wave-function makes sufficient overlap with a wave-function of the *P* state we can observe reaction from *R* to *P* even at the absolute zero of temperature. Furthermore, the rate of the reaction will be the same at all temperatures at which the bulk of the molecules are in their lowest vibrational state. This is the temperature-independent region.

As the temperature is raised and molecules go into higher vibrational states the rate increases because the higher vibrational wave-functions tunnel into a narrower width of barrier and have stronger overlap with those from *P*. This is the region of temperature-dependent nuclear tunnelling.

The strongest overlap occurs at the level at which the *R* and *P* curves cross, i.e. at the top of the barrier. If the temperature is high enough that a sufficient fraction of molecules are excited to this level, one finds that most of the reaction will occur through this vibrational level and we observe the region of Arrhenius dependence on temperature (equations (3.1) or (3.2)). An example showing the three regions is worked out in detail in Section 5.3.1. Plots of temperature dependence typically have the appearance of curves A and B in Fig. 1.1.

A second characteristic of nuclear tunnelling is the possibility of strong isotope effects. The penetration of the barrier is given by equation (2.1) or, for different barrier shapes, by expressions with similar exponentials such as equation (5.97). In all cases the square root of the mass of the penetrating particle is a factor in the exponent, and thus gives an exponential dependence on mass.

Here we briefly review the place of nuclear tunnelling in chemistry. The first recognized case, 'umbrella inversion' of the NH_3 molecule, was studied by Hund (1927) in connexion with the existence of stereo-isomerism in the light of the quantum-mechanical possibility of penetration of barriers to rearrangement of molecular configurations. He concluded that the exponential behaviour of tunnelling rates allowed such a wide variation that both the practically infinite lifetime of stereo-isomers and also molecules that resonate rapidly between isomeric structures can all be accommodated.

The suggestion by Bourgin (1929) that the decomposition of N_2O_5 involves tunnelling has also been mentioned.

Bell (1933, 1935) considered atomic tunnelling for some chemical reactions. His model was that of a particle attacking a fixed one-dimensional barrier with a Boltzmann distribution of kinetic energies. He used both a barrier designed by Eckart (1930) for which the Schroedinger equation can be solved exactly and (Bell, 1935) a parabolic barrier. The Eckart barrier is described later in Section 5.6 (Fig. 5.9). Bell concluded that tunnelling of protons should be quite common. Further discussion is given by Cribb, Nordholm & Hush (1979).

Bawn & Ogden (1934) included tunnelling in their discussion of the reactions of hydrogen isotopes.

Christov (1958, 1960) suggested that light ions (H^+, Li^+, Be^{2+}) approaching an electrode may tunnel a barrier between a state not-bound to the electrode and a bound (adsorbed) state.

As mentioned earlier, Sutin (1962) and Newton (Sutin, 1979) examined the barrier to the shift of water molecules in the hydration sphere of Fe^{2+} ions on changing to Fe^{3+} and also the Franck–Condon factors and concluded that the water might tunnel this barrier during electron exchange. Levich (1966) also concerned himself with atomic tunnelling as part of the total process. This aspect will be considered in more detail in the section reviewing theoretical developments.

Child (1967) investigated the effect of a dip in the top of a reaction barrier, i.e. a double-humped barrier, and showed that at low temperatures the reactants would tend to tunnel the humps to the dip, thus making the activation energy less at low temperatures than at high temperatures.

Willard and co-workers (Neiss, Sprague & Willard, 1975; Wilkey & Willard, 1976) provide experimental evidence of tunnelling in extraction of H atoms from molecules by free radicals.

Perhaps the first experimental evidence for chemical reaction by nuclear tunnelling of something larger than protons is the observation of the polymerization of formaldehyde to polyoxymethylene by Goldanskii and co-workers (1973, 1976, 1979*a*, *b*). A number of groups, e.g. Bruk *et al.* (1966) and Barkalov *et al.* (1966) (see Goldanskii, 1975) had observed yields of polymers induced by irradiating various organic monomers with ionizing radiation at low temperatures. The yields, however, had been determined after warming again to room temperature. In order to determine whether or not the polymerization actually occurred at the low temperature, Goldanskii's group developed sensitive calorimeters with short time constants. For example, at 4 K the calorimeter time constant was 2 to 3 seconds while the heat evolution from the polymerization decayed with a half-life of some 10 seconds after the ionizing radiation used to initiate the reaction was switched off. From the amount of heat evolved and the measured heat of polymerization of formaldehyde they calculate chain lengths of 10^3 and a mean time for addition of a link to the polymer chain of about 10^{-2} s. The rate is constant up to a temperature of about 10 K and then rises to give a mean time per link of about 10^{-5} s at 80 K. Experiments with deuterated formaldehyde showed no great isotope effect and they conclude that the tunnelling is of whole formaldehyde molecules rather than of hydrogen atoms.

The process is pictured in Fig. 3.3 A. The distance between neighbouring formaldehyde molecules before polymerization is determined by the Van der Waals forces and is larger by about 0.4–0.5 Å than the bonded distance in the polymer. Between the time of initiation by the radiation and termination by some reaction that 'caps' the end of the chain, the growing end of the chain will have either a charge or an unpaired electron which invites the polymerization. There is a barrier, however, as shown by the 10 ms time required for reaction below 10 K and by the 0.1 eV activation energy at higher temperature. The reaction consists, therefore, of the neighbouring molecule of formaldehyde tunnelling the barrier to arrive at bonding distance from the end of the chain. It is unclear whether or not an electron might also tunnel in the process but in any case the electrons move to cement the new bond and expose the new terminus of the chain for the next step. Goldanskii (1979*a*) discusses the mechanism within the framework of the quantum-mechanical theory of non-radiative transitions (see chapter 5).

There has been some discussion (Wickramasinghe, 1974; Fawcett, 1975; Wickramasinghe & Santhanan, 1975) over possible implication of formaldehyde polymerization in cosmology. Goldanskii (1979*a*) points out that the possibility of chemical reaction near absolute zero removes any restrictions due to decrease of entropy and only requires that the reaction be exothermic. Thus it is possible that in prebiotic conditions polymers of importance to biology such as polypeptides and nucleic acids could have formed at temperatures near absolute zero.

More recently Goldanskii's group (Kiryukhin, Barkalov & Goldanskii, 1979; Barkalov, *et al.*, 1980) has studied the combination of ethylene with

Fig. 3.3. Tunnelling processes observed at low temperature by Goldanskii and co-workers. (A) Polymerization of formaldehyde. (B) Hydrobromination of ethylene.

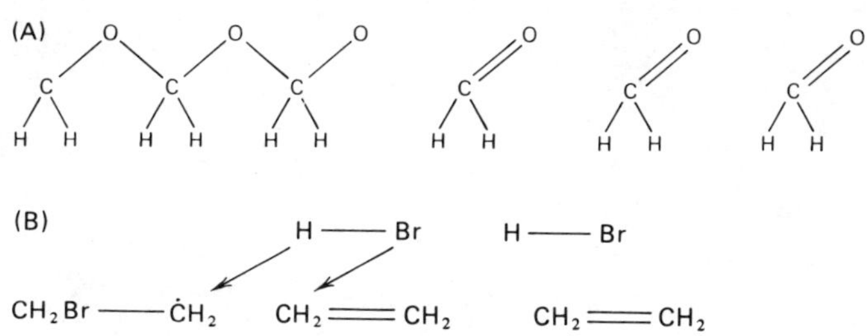

hydrogen bromide at temperatures between 40 and 90 K, initiated by ionizing radiation. They find a chain-reaction mechanism started by bromine atoms split from HBr, as follows:

$$\dot{Br} + C_2H_4 \rightarrow \dot{C}_2H_4Br,$$

$$HBr + \dot{C}_2H_4Br \rightarrow C_2H_5Br + \dot{Br}.$$

The time per single step increases from near 10^{-5} s at 90 K to 0.1 s at 70 K with activation energy equal to 0.17 eV. Below 70 K it is constant at 0.1 s. Deuteration of either compound changes the rates by only small amounts. It is concluded that the rate is determined by the tunnelling of HBr or Br and C_2H_4 toward each other as shown in Fig. 3.3 B and not by the hydrogen motion. A discussion of mechanism for reactions in the solid phase at low temperatures is given by Benderskii, Goldanskii & Ovchinnikov (1980).

4

Tunnelling in biology

This chapter will review the available data pertinent to the occurrence of tunnelling in biological systems. Most of it will be the temperature-dependences of the rates of reaction. As already mentioned, electron transfer has two aspects. One is the electron-transfer probability when the nuclei are in the correct position to satisfy both the Franck–Condon principle and conservation of energy. This is handled by κ in equation (3.1). The other is the nuclear motions, both caused by the changed polarities resulting from the electron transfer and necessary to provide an activated, non-equilibrium polarity which allows electron transfer without change of energy. Temperature dependence results from the activation necessary for the nuclear motion, and shows that the nuclear motion is necessary to enable the electron transfer. If the electron-transfer reaction shows temperature dependence at high temperature and temperature independence at low temperature (Fig. 1.1), this can be taken as good evidence that the nuclei are tunnelling their activation barrier at the low temperatures.

Evidence for electron tunnelling resides in the value of κ. If κ is appreciably less than 1, then electron tunnelling is slowing the reaction. A clue to the value of κ is the value of k_∞, equation (3.2). Equations (3.1) and (3.2) assume that no nuclear tunnelling is occurring. In such a case, equations (3.1) and (3.2) show that

$$k_\infty = \kappa B \exp(\Delta S^\ddagger / k_B). \tag{4.1}$$

If κ is less than 1, then k_∞ is less than $B \exp(\Delta S^\ddagger / k_B)$. Under equation (3.2) it is stated that B is about $10^{13}\ s^{-1}$ for intramolecular electron transfer and about $10^{11}\ \text{M}^{-1}\ s^{-1}$ for bimolecular. If the entropy of activation $\Delta S^\ddagger$ can be estimated κ can be calculated from k_∞. In many cases

$|\Delta S^{\ddagger}|$ can be considered small enough to ignore and then the criterion for electron tunnelling becomes: $k_{\infty} \ll 10^{13}\,\mathrm{s}^{-1}$ for the monomolecular case and $k_{\infty} \ll 10^{11}\mathrm{M}^{-1}\mathrm{s}^{-1}$ for the bimolecular case.

In presenting the biological data we will calculate the apparent k_{∞} values given by the data, but the reader will have to use his own judgement as to applicability.

One can be quite sure that electron transfer is by electron tunnelling if it is known that there is a barrier between donor and acceptor more than, say 5 Å thick, which is either vacuum or contains material whose electron affinity is appreciably less than that of the original donor.

All of these criteria will be discussed more thoroughly in Chapter 5.

4.1 Photosynthesis

For those not familiar with photosynthesis, a brief explanation may be in order. The current picture of the bacterial photosynthetic 'reaction centre' (RC) consists of (1) a special dimer of bacteriochlorophyll

Fig. 4.1. The scheme of photosynthesis in bacteria. On the left is shown the electron transport paths as found in the green bacteria, on the right as in the purple bacteria. Both extract electrons from substrate and drive them into the CO_2 reduction cycle. Both also use cyclic electron flow to phosphorylate ADP to ATP. From DeVault (1979), p. 306.

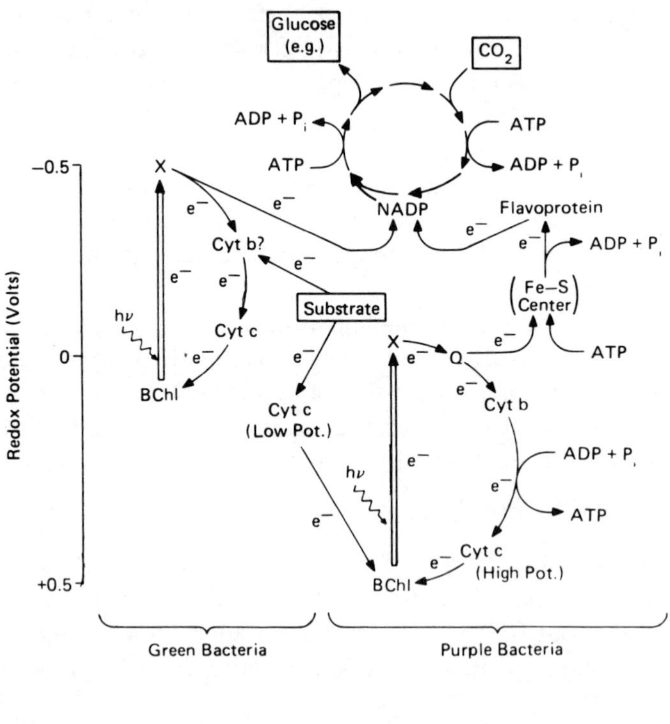

(BChl), the 'primary (electron) donor', symbol P (sometimes written with one of the main absorption peaks, in nm, as 'P-870'); (2) a bacteriopheophytin molecule, symbol I, (for 'intermediate' primary electron-acceptor); (3) a 'primary acceptor', probably an iron–quinone complex, symbol A; (4) one or more quinones as secondary acceptors: Q; (5) one or more cytochromes as secondary donors C; (6) other molecules not relevant to our discussions. In the resting state, the system may be written: $C \cdot P \cdot I \cdot A \cdot Q$. In the normal course of events, an excitation resulting from photon absorption in the antenna pigments is funnelled to the RC BChl dimer: $C \cdot P^* \cdot I \cdot A \cdot Q$. Within 10 ps an electron transfers from P to I: $C \cdot P^+ \cdot I^- \cdot A \cdot Q$ (Kaufmann *et al.*, 1975, 1976; Rockley *et. al.*, 1975; Dutton *et al.*, 1975; Moscowitz & Malley, 1978). In 140 ps ($t_{\frac{1}{2}}$) the electron goes to A: $C \cdot P^+ \cdot A^- \cdot Q$ (Kaufmann *et al.*, 1975; Rockley *et al.*, 1975; Holten *et al.*, 1978*b*). In 1 μs in *C. vinosum* (DeVault & Chance, 1966; Parson, 1968; Seibert & DeVault, 1970, 1971: note revision from earlier value of 2 μs) or sooner or later in other species (Kihara & Chance, 1969), the cytochrome is oxidized: $C^+ \cdot P \cdot I \cdot A^- \cdot Q$. This is the reaction whose kinetics are plotted in curves A and B of Fig. 1.1. Within 60 μs to 1 ms (Parson, 1969*a*) the electron moves from A on to Q: $C^+ \cdot I \cdot A \cdot Q^-$. The result is a relatively stable charge-separation which is now able to furnish energy (i) for phosphorylation (in a cyclic electron transport) or (ii) for CO_2 reduction, which takes electrons from substrate and uses them to reduce NAD which can then reduce compounds formed with CO_2. These paths are shown in Fig. 4.1, where electron transfers are indicated by the symbol e^- on the arrows. The green plants

Fig. 4.2. Photosynthesis as found in green plants. Electrons are extracted from water by PS-II and driven to the CO_2 reduction cycle by PS-I. From DeVault (1979), p. 308.

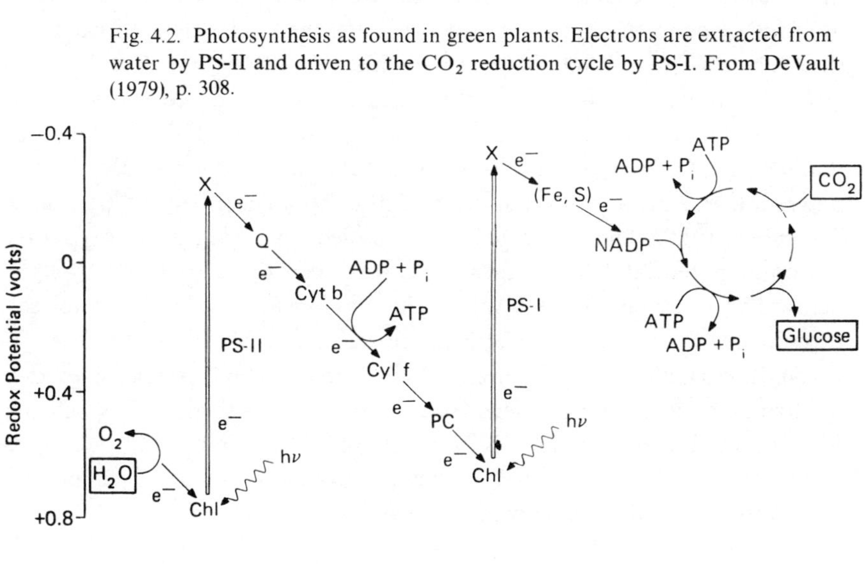

have two such systems whose details are less well known. Their photosystem I (PS-I) is probably quite similar to the bacterial and serves the same end of driving electrons to CO_2 reduction. P-700 is the name given the oxidizable chlorophyll of its RC. The green-plant photosystem II (PS-II), whose oxidizable chlorophyll is called P-680, extracts electrons from water, producing O_2, a function which the bacteria do not have. The electrons from PS-II go to PS-I in several steps. The complete scheme is shown in Fig. 4.2. References for more details are: Govindjee (1975, 1982) and Barber (1977).

In the bacterial system, if for any reason the cytochrome oxidation cannot occur and reoxidation of Q^- by external agents is blocked, it is possible for the 'reverse primary reaction' to occur:

$$P^+ \cdot I \cdot A \cdot Q^- \rightarrow P^+ \cdot I \cdot A^- \cdot Q \rightarrow P \cdot I \cdot A \cdot Q.$$

This takes 10–100 ms and is the reaction plotted in curves B and C of Fig. 1.1. If the formation of Q^- has been prevented in any way then, of course, we have only the second part in the reverse reaction. If the formation of A^- has been prevented, the state $P^+ \cdot I^-$ can live for 10–30 ns (Parson, Clayton & Cogdell, 1975) during which the electron spins on the two radicals can precess in and out of phase, giving alternate singlet and triplet states (Dutton *et al.*, 1972; Thurnauer, Katz & Norris, 1975; Prince, Leigh & Dutton, 1976; Blankenship, Schaafsma & Parson, 1977; Hoff *et al.*, 1977) and decay to ground state via one or the other, or by passing triplet excitation to carotenoids.

To understand the subtleties in this work one needs to know some of the species differences. In *C. vinosum* (formerly *Chromatium D*) there are four cytochrome molecules (of two different kinds) close to each RC (Bartsch & Kamen, 1960; Olson & Chance, 1960; Parson, 1969*b*; Dutton, 1971; Dutton *et al.*, 1976), and they are tightly bound to the RC in a hydrophobic environment (Cusanovich & Bartsch, 1969; Kennel & Kamen, 1971). If one tries to separate RC from antenna BChl with detergent, the cytochromes stick with the RC (Thornber, 1970; Lin & Thornber, 1975). The two cytochrome molecules which have low redox midpoint potential are the ones which are seen to be oxidized by light below 150 K (Seibert & DeVault, 1970; Parson & Case, 1970). After the cytochrome is oxidized there is no re-reduction of it below 200 K in *C. vinosum* (DeVault & Chance, 1966). The two high-potential (better oxidizers) cytochromes are seen to donate electrons to the RC when the low-potential cytochromes are already oxidized (Seibert & DeVault, 1970; Parson & Case, 1970), but they are unable to do so in *C. vinosum* below about 180 K (Vredenberg & Duysens, 1964).

The cytochromes in *Rhodopseudomonas sphaeroides* are more loosely bound. They fall off when treated by detergent in the preparation of purified RC s. These RC preparations (Reed & Clayton, 1968; Clayton & Wang, 1971; Feher, 1971) may also be missing the Q. However, they perform the reactions leading to $P^+ \cdot I \cdot A^-$ as efficiently as in *vivo* and also the return reaction to $P \cdot I \cdot A$. The Q and the C can be added back and are then fully functional. *R. sphaeroides* (Kihara & Chance, 1969) does not show cytochrome oxidation at 77 K. This temperature also freezes out the reduction of Q, but not the primary reaction (to $P^+ \cdot I \cdot A^-$) or its reverse. *Rhodospirillum rubrum* (Chance & Nishimura, 1960) and *Rhodopseudomonas capsulata* also do not show cytochrome oxidation at 77 K (Kihara & Chance, 1969).

'Chromatophores' often used in research on the bacterial systems are a cell-free extract from photosynthetic bacteria retaining nearly all the membranous constituents not soluble in detergentless buffer. Under proper conditions they can do most of the reactions of photosynthesis, are more accessible to reagents, and do not strongly scatter photometric light beams.

'Chloroplasts' are the organelles in the cells of green plants and algae that contain the whole photosynthetic apparatus. For research purposes they are often separated from the rest of the cell constituents.

Subchromatophore and subchloroplast fragments are prepared from chromatophores and chloroplasts, respectively, by treatment with detergents. Usually the effort is to get rid of antenna chlorophyll and any other parts not directly involved in the early reactions of photosynthesis. The reaction-centre preparations made from *R. sphaeroides* by Clayton and co-workers and the subchromatophore preparation from *C. vinosum* by Thornber were described above. From chloroplasts it is possible to separate particles containing PS-I reaction centres from those containing PS-II (Boardman & Anderson, 1964; Vernon & Shaw, 1971).

4.1.1 *Photosynthetic reaction-centre–acceptor reactions*

Table 4.1 is a summary of data on the temperature dependence of electron-transfer reaction between the reaction centre and several early acceptors of the electron. Most of them concern the rate of return of the electron to P^+, which occurs only when the normal secondary reactions are somehow prevented. The first column designates the species and the type of preparation. The second column indicates the particular electron-transfer reaction. The abbreviations are explained below the table. The next three columns apply to the room temperature range and give the half-time at 298 K, the Arrhenius pre-exponential factor k_∞ and the Arrhenius

Table 4.1. *Photosynthetic reaction-centre–acceptor reactions*

Species	Electron transfer	High-temperature range $t_{\frac{1}{2}}$(298 K) (ms)	k_{∞} (s^{-1})	$E^{\ddagger}$ (eV)	T_x (K)	Low-temperature range $t_{\frac{1}{2}}$(80 K) (ms)	k_{∞} (s^{-1})	$E^{\ddagger}$ (eV)	References
Bacteria									
Rsp. rubrum	$P^+ \leftarrow A^-$	51.8	4.4	− 0.029	162	17.0	29	− 0.0024	(*a*)
	$P^+ \leftarrow A^-$	101	1.47	− 0.040	138	16.6	42	0	(*b*)
(WC)	$P^+ \leftarrow A^-$	—	—	—	—	(16.3)	42.6	0	(*c*: table 2)
($C\phi$)	$P^+ \leftarrow A^-$	{22	—	—	—	6.8	—	—}	(*d*)[†]
(extracted)		{59	—	—	—	19	—	—}	
Rps. sphaeroides									
(WC)	$P^+ \leftarrow A^-$	—	—	—	—	(21.1)	32.8	0	(*c*: table 2)
(RC)	$P^+ \leftarrow A^-$	—	—	—	—	(18.6)	37.3	0	(*c*: fig. 3)
	$P^+ \leftarrow A^-$	74.3	1.37	− 0.049	173	18.6	37	0	(*e*)
	$P^+ \leftarrow A^-$	{(242)	0.074	− 0.094	—	—	—	—}	(*f*)[‡]
		{91	6.81	− 0.003	234[‡]	—	—	—}	
	$P^+ \leftarrow A^-$	{(331)	0.049	− 0.096	173	21.8	31.8	0}	(*g*)[‡]
		{181	2.49	− 0.011	253[‡]	—	—	—}	
	$P^+ \leftarrow A^-$	90	1	− 0.052	165	17	36	− 0.001	(*h*)
	$P^+ \leftarrow A^-$	82	1.8	− 0.040	—	—	—	—	(*i*)
	$P^+ \rightarrow I$	$< 6 \times 10^{-9}$	—	—	—	$< 6 \times 10^{-9}$	—	—	(*j*)
	$I^- \rightarrow A$	1.0×10^{-7}	6.7×10^{9}	0	None	1.0×10^{-7}	6.7×10^{9}	0	(*j*)
C. Vinosum ($C\phi$)	$A^- \rightarrow Q$	0.043	1.9×10^{10}	+ 0.36	—	—	—	—	(*k*) (2nd order?)
Green plant, PS-I									
Spinach (Clp)	$P^+ \leftarrow A^-$	—	—	—	—	{0.122	5.6×10^{3}	0}	(*r*)[†]
						{1.7	4.1×10^{2}	0}	
Spinach (Clp)	$P^+ \leftarrow A^-$	10.9	6×10^{4}	+ 0.176	{198	1350	2.0	0}	(*l*)[†]
					{141	23×10^{3}	0.03	0}	

PS-I particles	$P^+ \leftarrow A^-$	$\{44 \times 10^3$	7×10^8	$+0.63$	240‡	—	—	— }	(*m*)
		$(14 \times 10^3 - 0.4)$	$6.7 - (2.4 \times 10^5)$	$+0.13$	—	—	—	— }	
PS-I particles	$P^+ \leftarrow A^-$	0.021	1.3×10^5	$+0.035$	89	0.52	1.31×10^3	0	(*n*: fig. 5)
PS-I particles	$P^+ \leftarrow I^-$	0.028	1.2×10^4	$+0.041$	64	(1.0)	7×10^2	0	(*o*: fig. 4)
PS-I particles	$P^+ \leftarrow A^-$	—	—	—	—	(130)	—	—	(*o*)
Green Plant, PS-II									
Spinach (Clp)	$P^+ \leftarrow A^-$	(53)	5×10^6	$+0.3$	—	—	—	—	(*p*: fig. 1)
Bean (Clp)	$A^- \rightarrow Q$	$\{0.064$	1.5×10^8	$+0.24$	290‡	—	—	— }	(*q*)‡
		(0.063)	7.9×10^6	$+0.17$	—	—	—	— }	
Lettuce (Clp)	$A^- \rightarrow Q$	0.086	1.4×10^7	$+0.19$	—	—	—	—	(*q*)
Pea	$A^- \rightarrow Q$	0.049	2.3×10^7	$+0.19$	—	—	—	—	(*q*)
Pea + Grm (Clp)	$A^- \rightarrow Q$	0.057	1.1×10^7	$+0.18$	—	—	—	—	(*q*)
Spinach (Clp)	$A^- \rightarrow Q$	$\{0.047$	5.4×10^8	$+0.27$	290‡	—	—	— }	(*q*)‡
		(0.039)	3.0×10^9	$+0.31$	—	—	—	— }	
Spinach +	$A^- \rightarrow Q$	{(0.055)	2.2×10^9	$+0.31$	290‡	—	—	— }	(*q*)‡
Glut. (Clp)	$A^- \rightarrow Q$	(0.044)	1.9×10^9	$+0.30$	—	—	—	— }	

( ) indicate value of $t_{\frac{1}{2}}$ extrapolated (by Arrhenius law) to a temperature outside range of measurement or validity.
‡ Two high-temperature regimes with break between them at T. † Two kinetic phases at same temperature.

Abbreviations

T_x = temperature of intersection of high- and low-temperature asymptotes on an Arrhenius plot. k_∞ = Arrhenius pre-exponential factor. $E^\ddagger$ = Arrhenius 'activation energy'. Negative means reaction is faster at lower temperature. *Rsp* = *Rhodospirillum*. *Rps* = *Rhodopseudomonas*. WC = whole cells. RC = reaction centre preparation. Cϕ = chromatophores. Clp = Chloroplasts. Grm = gramicidin. Glut = glutaraldehyde.

References

(*a*) Parson (1967); (*b*) Loach *et al.* (1975); (*c*) McElroy *et al.* (1974); (*d*) Morrison & Loach (1978) (see text); (*e*) Hsi & Bolton (1974); (*f*) Clayton & Yau (1972); (*g*) Romijn & Amesz (1976); (*h*) Clayton (1979); (*i*) Noks *et al.* (1977); (*j*) Avouris *et al.* (1978); (*k*) Parson (1969*a*) analysed as first order, may be second order; (*l*) Visser *et al.* (1974); (*m*) Ke *et al.* (1979*a*, *b*); (*n*) Mathis *et al.* (1978); (*o*) Shuvalov *et al.* (1979); (*p*) Knaff & Malkin (1974); (*q*) Jursinic & Govindjee (1977) measured by delayed light emission; (*r*) Mathis & Conjeaud (1979).

activation energy, $E^{\ddagger}$, respectively (see equation (3.2)). A negative value for the latter simply means that the reaction is slower at higher temperature. Columns 7–9 give the similar data for the low-temperature range, generally below 100 K. The half-time is given for 80 K. If a half-time at the designated temperature had to be obtained by extrapolation outside the range of measurement or of validity it is enclosed in parentheses. The sixth column, designated 'T_x', gives the temperature at which the high- and low-temperature asymptotes on an Arrhenius plot intersect. It can be calculated by: $T_x = (E_1{}^{\ddagger} - E_2{}^{\ddagger})/k_B \ln(k_{\infty 1}/k_{\infty 2})$, where $E_1{}^{\ddagger}$, $E_1{}^{\ddagger}$, $k_{\infty 1}$ and $k_{\infty 2}$ come from columns 5, 9, 4 and 8, respectively. Some of the data in Table 4.1 (and Table 4.2 and 4.3) are measured and calculated from authors' figures, and don't always agree with values in their texts.

One notices that, as pointed out by Hales (1976), the bacterial systems all have negative temperature coefficients for the reverse primary reaction while those of the green plant systems are positive in the high-temperature region. The cyanobacteria (also called blue-green algae), although they are prokaryotes (do not have nuclei) like bacteria, they do have the green-plant type of photosynthetic system, Fig. 4.2, and positive temperature coefficients for the reverse primary reactions (data not shown in Table 4.1).

There is a warning not to take the data of Tables 4.1–4.3 too literally. In much of the rather preliminary type work done so far, it has not always been easy to distinguish exponential time courses from other rate laws or even biphasicity. It is also not always clear that the rates observed apply to the reaction as given and not to some rate-limiting intermediate step. Clayton (1979) has indicated biphasicity in his observations and presents the fraction of each phase. The part presented in Table 4.1 (ref. *h*) is the 'fast phase' as observed in aqueous suspension of RCs. It is 85–96% of the total decay of P^+. He also obtained somewhat similar data from air-dried films and dehydrated films of RCs.

Morrison & Loach (1978) (ref. *d* in Table 4.1) worked with preparations from which the quinone acceptors had been extracted by organic solvents down to a level of 0.5 quinone molecules per RC. They showed that the biphasicity they observed in the reversed primary reaction is quantitatively explained by the assumption that one quinone serves two reaction centres so that the two phases could be represented by:

$$P^+ \cdot Q^= \cdot P^+ \rightarrow P \cdot Q^- \cdot P^+ \rightarrow P \cdot Q \cdot P.$$

Lukashev *et al.* (1976) working with *Rsp. rubrum* and *Ectothiorhodospira shaposhnikovii* showed that the character of the biphasicity of the reversed primary reaction in these bacteria depended upon whether the samples were frozen in the light or in the dark. A fast phase (about 50 ms) appeared at low temperatures (< 230 K) if frozen in the dark, but not otherwise. The total amplitude of re-reduction of P^+ diminished at low temperature if frozen in the light. They interpret this to mean that the lower the temperature the more the secondary acceptor (Q) becomes disconnected from the primary acceptor (A).

Ke *et al.* (1979*a*, *b*), working with subchloroplast fragments containing only PS-I prepared from spinach, present quite complicated data on the recovery rates of P-700 after its photo-oxidation. These fragments do not contain cytochrome (or plasto-cyanin which replaces cytochrome as secondary donor in PS-I and so the recovery of P-700 is the reverse primary reaction (Vernon & Shaw, 1971). Their data, partly reproduced in Table 4.1 (ref. *m*), is an 80–100% component of the P-700 recovery in the range 294–240 K. At the latter temperature it becomes too slow to observe but at 220 K is suddenly replaced by a 90% fast recovery: $t_{\frac{1}{2}}$ for slowest part is about 6 s. As the system is further cooled the amplitude of the fast recovery decreases to 17% at 20 K, and there is then very little measurable slow recovery either. They analyse the recovery kinetics below 220 K as multiphasic with a distribution of values of k_∞, giving an overall decay that is approximately linear on a logarithmic time scale. The slopes of the lines on such a plot do not vary with temperature and they show that this means a distribution of the values of k_∞ (or of the entropy of activation, $S^\ddagger$).

Sauer *et al.* (1978) propose a series of four acceptors from P-700, the primary donor of PS-I. They name the first two 'A_1' and 'A_2'. The third and fourth are 'centres A and B of P-430' (Hiyama & Ke, 1971; Malkin & Bearden, 1971; Evans, Reeves & Cammack, 1974; Evans, Sihra & Cammack, 1976; Ke, Hansen & Beinert, 1973). Shuvalov, Ke & Dolan (1979, ref. (*o*) in Table 4.1) propose that A_1 is a chlorophyll *a* dimer while the others are thought to be iron–sulphur centres on the basis of EPR spectra. Mathis & Conjeaud (1979) studied the temperature dependence at low temperatures of the reverse reaction in PS-I. Intact chloroplasts showed a 122 μs ($t_{\frac{1}{2}}$) phase, absent from more degraded preparations. Floyd, Chance & DeVault (1971) had earlier reported a 30 μs reduction of P-700 at low temperature. In Table 4.1 A_1 and A_2 are designated 'I' and 'A' respectively to be consistent with the symbols used in this book for the same functions in bacterial centres.

In PS-II the acceptor, which is called 'A' for internal consistency in this book, is usually called 'Q'. Thus in reporting the work of Jursinic & Govindjee (1977) (see also that of Knaff & Malkin, 1974, ref. (*p*) in Table 4.1) their 'Q_1' and 'Q_2' are called here 'A' and 'Q'. Although delayed light emission is thought to be associated with the reverse primary reaction in PS-II the rate of its decay reported in Table 4.1 (ref. (*q*)) is thought to be controlled by the faster electron movement from Q_1^- to $Q_2(A^- \rightarrow Q)$.

Itoh (1978) (Matsuura *et al.*, 1979) has studied the effects of the electric fields of the ionic double layer at the surface of the thylakoid membranes (of chloroplasts) on rates of electron transport between the acceptors in PS-II and artificial acceptors in aqueous solution. The dependence of rate on ionic strength, electric charge of acceptor, pH and detergents all agreed with the idea that the rate depended on the concentration of external acceptor and other field effects which can be calculated from the charge of the double layer. The field effects of surface potentials on P-700 reduction was studied by Tamura, Yamamoto & Nishimura (1980).

The temperature dependence of the rate of electron transfer from I^- to A was studied experimentally and theoretically by Schenck *et al.* (1981) in reaction-centre preparations. By measuring at smaller temperature intervals than had been done before they discovered a sharp maximum of rate at about 25 K. This could be fitted theoretically by assuming two soft modes of vibration coupled to the electron transfer (see Section 5.4.1). Upon deuteration the sharp maximum in the rate disappears and is replaced by a broad plateau from 200 to 4 K, faster than the room-temperature rate. This was only partially fitted by theory.

(*a*) *Slowness of reversed rate.* The near 100% quantum efficiency of photosynthetic electron transport depends upon the slowness of the rate of return of the electron from the various acceptors to P^+. Hopfield (1976), Blankenship & Parson (1979) and Beitz & Miller (1979*b*), for example, discuss the problem. *In vitro* experiments on model systems invariably find reverse rates as fast or faster than the forward rates (Blankenship & Parson, 1979; Huppert, Rentzepis & Tollin, 1976; Holten *et al.* 1976, 1978*a*). The question is how does the photosynthetic system make the return rates so slow?

One suggestion is that the energy-gap law, equation (3.3), is used. The return reactions have a much larger $-\Delta G^\circ$ than the forward reactions assuming that the return is to the ground state of P. If λ is the same for both directions and equal to $-\Delta G^\circ$ for the forward direction, one gets maximum rate in the forward direction and, according to equation (3.3) a much reduced reverse rate. The problems concerning the energy-gap law were discussed earlier in Section 3.5.1. It appears that the photosynthetic system may have 'engineered' the vibronic couplings so that the energy-gap law is obeyed.

One could also wonder whether the matrix element, $|H_{ab}|$, for electron transfer between donor (P^* or I^-) and acceptor might be appreciably different from that between ground state P and acceptor. This possibility is considered by both Jortner (1980) and Warshel (1980). They both take an overall view which includes considering the high speed of the forward reactions as well as the slowness of the reverse. Jortner concludes that $P^*I \rightarrow P^+I^-$ is ultrafast with $|H_{ab}| = 10\,\text{cm}^{-1}$ (1.2 meV), the electronic transition competing directly with vibrational relaxation from a level of photo-excitation above the crossing point and occurring when the relaxation reaches the level of the crossing point and, therefore, good vibrational overlap; that $I^-A \rightarrow IA^-$ is fast because activationless $(-\Delta G^\circ = \lambda)$ and sets $|H_{ab}| = 4\,\text{cm}^{-1}$; that $P^+IA^- \rightarrow PIA$ is slow because $|H_{ab}|$ for the long distance involved is small $(5 \times 10^{-4}\,\text{cm}^{-1})$; and that $P^* \rightarrow P$ and $P^+I^- \rightarrow PI$ are blocked by large energy gaps.

Warshel (1980) calculates energy surfaces for the primary steps in bacterial photosynthesis. These are plots of energy v. nuclear coordinates like Fig. 3.1 but using more than one nuclear coordinate. He concludes that $|H_{ab}|$ for $P^*I \rightarrow P^+I^-$ results from overlap of π^* orbitals of both reactant and product and is thus large compared to the vanishing overlap of the π^* orbital in P^+I^- and the pertinent π orbital in PI which makes $|H_{ab}|$ for $P^+I^- \rightarrow PI$ very small. He also concludes that $P^+I^- \rightarrow PI$ has high activation energy as a result of large energy gap relative to small reorganization energy. He suggests that a fixed negative charge on the side of P opposite to that of I could have the double effect of speeding the movement of the electron from P to I and of attracting P^+ away from I^-, thus making the tunnelling path for return of the electron longer and $|H_{ab}|$ smaller than for the forward reaction.

Vorotyntsev & Itskovich (1980) attempt a general theory of charge separation and conclude that two initial steps are necessary to avoid fast reverse reaction and that a slower conformational change in position and orientation of products is needed to stabilize them.

(*b*) *Triplet states and magnetic fields.* A spectacular feature of photosynthesis is that the forward reactions take place with high yield in the singlet state in competition with fluorescence and radiationless decay which would return the system to its ground state and with intersystem crossing to triplet states through which most photochemical reactions proceed. The latter is believed to result from the alternation of singlet and triplet states to be found in the state P^+I^- as the two electron spins precess at different rates as mentioned earlier. This phenomenon is deduced from the effect of magnetic fields on rates of triplet state production (Blankenship, Schaafsma & Parson, 1977; Hoff *et al.*, 1977; Thurnauer, Katz & Norris, 1975). Hoff (1981) has written an excellent review of these effects and a clear explanation of the quantum mechanics involved. Norris *et al.* (1982) further characterize the P^+I^- state experimentally using triplet yield detection of magnetic resonance and they evaluate the parameters with further theory. Schenck, Blankenship & Parson (1982) measure the rate of decay of the state P^+I^- (called by them 'P^F') and of delayed fluorescence (which results from $P+I^- \rightarrow P^*I$) and the yields of these and of triplet *BChl* (called by them 'P^R') as affected by temperature, magnetic field, deuterium substitution, redox potential, depletion of quinone or iron and the presence of *o*-phenanthroline. They compare with the theories and conclude: 'Although the usual scheme seems satisfactory in outline the known decay paths of P^F cannot fully account for the dependence of the P^R yield and the decay kinetics of P^F on temperature and magnetic field nor for the decay kinetics and temperature dependence of the delayed fluorescence.'

Sonneveld (1981) reports extensive studies of fluorescence in photosynthesis. These include effects of magnetic fields. In PS-II (Sonneveld, Duysens & Moerdijk, 1980) he and collaborators find two components of delayed fluorescence above 200 K, one decaying in 700 ns and the other in about 150 ns. The latter persists down to 4 K and increases at the lower temperature and is interpreted to indicate the presence of an intermediate in electron transport, called '*W*' by him, corresponding to '*I*' in this book. It is increased by magnetic fields. The weak fluorescence from PS-I is also increased slightly by a magnetic field. It gives delayed fluorescence (indicating reversed electron transfer to give P^*) only at low redox potential.

The possibility that low-frequency vibronic modes play a part here as discussed in Chapter 5, should also be considered.

4.1.2 *Photosynthetic reaction-centre–donor reactions*

Data on electron transfer on the donor side of the reaction centre is collected in Table 4.2. The structure of the table is the same as that of Table 4.1 except that column 2 of that table is omitted. Instead we note that all reactions listed except the last five lines are transfers from a secondary donor to the primary donor of the reaction centre. The secondary donors are cytochromes except '*Z*' and possibly 'D_2' found in PS-II and plastocyanin in PS-I. In the data of Parson (1967, ref. (*j*) in Table 4.2) on *rubrum* the added PMS donates directly to the RC in a second-order reaction. The last five lines concern electron transfer to the secondary donor from some tertiary donor.

Altogether, Kihara & Chance (1969) found eleven species of photosynthetic bacteria which show cytochrome oxidation at 77 K. Quantitative data on temperature dependence of rate, however, are lacking for those not shown in Table 4.2 as also for *Ectothiorhodospira shaposhnikovii* studied extensively by Rubin and co-workers (Kononenko *et al.*, 1972). See also data on the re-reduction of cytochrome in *Rhodopseudomonas* sp. by Grigorov, Kononenko & Rubin (1970).

As in the case of *C. vinosum*, many, if not most, photosynthetic bacteria have both high- and low-potential cytochromes which can be photo-oxidized directly by the RC (Sybesma, 1969; Frenkel, 1970; R. Govindjee & Sybesma, 1972; R. Govindjee, Smith & Govindjee, 1974; Pyt'eva, Andreenko & Rubin, 1976; Pyt'eva, Ratyni & Rubin, 1976). Aeration ordinarily oxidizes the low-potential cytochrome and renders it unavailable for photo-oxidation. The cytochrome photo-oxidized under aerobic conditions can be expected to be a high-potential cytochrome. Kihara & Dutton (1970) observed that all the cytochromes photo-oxidized at 77 K under anaerobic conditions were irreversibly oxidized. Those photo-oxidized under aerobic conditions at 77 K (a less frequent occurrence) were reversible. These are indicated in Table 4.2 under the labels 'l.p.' for the former and 'h.p.' for the latter.

The delayed light decay rates of Jursinic & Govindjee (1977; ref. (*n*) in the time range shown in Table 4.2) correlate with fluorescence rise times which are believed to result from reduction of P^+ by the PS-II donor, '*Z*'.

The cytochrome oxidized in green leaves at 77 K (Chance & Bonner, 1963) turned out to be that called b_{559} (Knaff & Arnon, 1969; Floyd *et al.*, 1971) and the reaction centre oxidizing it turned out to be that of PS-II: P-680. The role of b_{559} in green-plant photosynthesis remains unclear despite considerable work by Butler and co-workers (Butler, Visser & Simons, 1973; Lozier & Butler, 1974), by Cramer & Whitmarsh (1979) (Whitmarsh & Cramer, 1977) and by Mathis and co-workers (Vermeglio & Mathis, 1973*a*, *b*; Mathis, Haveman & Yates, 1976). The apparent half-time for the photo-oxidation is 4.6 ms, approximately constant between 80 and 220 K (Floyd *et al.*, 1971; ref. (*k*) in Table 4.2), but there is no evidence that the reaction occurs at room temperature. Only about 15% is oxidized in a single flash (Floyd *et al.*, 1971) so that one can suppose that the reaction time has been shortened from, say, 30 ms by competition with reversed primary reaction (Butler, 1973). Cramer & Whitmarsh (1979) and Cramer, Whitmarsh & Horton (1979) argue that if electron-tunnelling distances must be small (< 10 Å), then b_{559} must be situated in the membrane close to P-680, and since b_{559} is accessible to water-soluble oxidants like ferricyanide, both b_{559} and P-680 must be located near the outer side of the vesicular membrane. This contradicts a postulate that P-680 is near the inner side of the membrane, resulting from consideration of proton movements.

Table 4.2. *Photosynthetic reaction-centre–donor reactions*

Species	High-temperature range			T_x (K)	Low-temperature range			References
	$t_{\frac{1}{2}}$(298 K) (μs)	k_∞ (s^{-1})	$E^\ddagger$ (eV)		$t_{\frac{1}{2}}$(80 K) (ms)	k_∞ (s^{-1})	$E^\ddagger$ (eV)	
Cytochrome photo-oxidation								
Bacteria								
C. vinosum (WC) (l.p.)	1.07	7.8×10^8	0.18	142	2.5	2.8×10^2	0	(*a*, *b*)
C. vinosum (WC) (h.p.)	2.13	2.6×10^7	0.11	×	×	×	×	(*b*, *c*)
C. vinosum (RC) (l.p.)	1.13	5.5×10^8	0.18	139	2.6	2.7×10^2	0	(*d*)
C. vinosum (l.p.)	1.4	8.5×10^8	0.19	150	2.3	3.1×10^2	0	(*e*)
Rps. sp. *N.W.* (l.p.)	0.55	4.0×10^7	0.091	187	0.0045	1.5×10^5	0	(*f*)
Rps. gelatinosa (h.p.)	0.70	1.0×10^6	0.10	201	0.0046	1.5×10^5	0	(*f*)
T. f loridana (l.p.)	1.80	—	—	—	0.6	—	—	(*f*, *g*)
Rps. viridis (h.p.)	< 1	—	—	—	0.010	—	—	(*h*)
Rps. palustris (h.p.)	1.8, 2.8	—	—	—	12, 7.3	—	—	(*i*, *h* (resp.))
Rsp. rubrum (Cϕ)(PMS)	470	3.2×10^{10}	0.43	—	—	—	—	(*j* : fig. 12)
Green plant, PS-II								
Spinach (Clp) (b_{559})	×	×	×	×	4.5	88	−0.004	(*k*)
Spinach (Clp) ('D_2')	—	40(at 80–100 K)	—	60–80	(630)	1.0	−0.0003	(*l*, *m*)
Pea (Clp) ('*Z*')	5 (ms)	1.7×10^5	0	—	—	—	—	(*n*)
Green plant, PS-I								
Spinach (Clp) (cytf)	2.9	2.7×10^{10}	0.48	×	×	×	×	(*p*)

Cytochrome reduction after photo-oxidation								
Bacteria								
	(s)				(s)			
C. vinosum (l.p.)	> 10	$< 2 \times 10^{10}$	0.67	×	×	×	×	(*c*)
C. vinosum (h.p.)	0.008	1.1×10^{11}	0.54	×	×	×	×	(*c*)
Rps. gelatinosa (h.p.)	0.21	6.4×10^{8}	0.49	—	—	—	—	(*f*)
	1.1	2.7×10^{7}	0.45	—	—	—	—	
Rps. palustris	—	—	—	—	10	1.6	0.02	(*g*, *o*)

( ) indicate value of $t_{\frac{1}{2}}$ extrapolated to a temperature outside the range of measurement or validity.

Abbreviations

T_x = temperature of intersection of high- and low-temperature asymptotes on an Arrhenius plot. k_∞ = Arrhenius pre-exponential factor. $E^\ddagger$ = Arrhenius 'activation energy' (negative means reaction is faster at lower temperature). C = *Chromatium*. *Rps.* = *Rhodopseudomonas*. *T* = *Thiocapsa*. WC = whole cells. RC = reaction centre or subchromatophore preparation. Cϕ = chromatophores. Clp = Chloroplast preparation. l.p. = low-potential cytochrome. h.p. = high-potential cytochrome. D_2: see ref. (*m*). *Z* = immediate donor to RC of PS-II, is not a cytochrome and probably contains manganese; see ref (*n*). × = no reaction observable. PMS = *N*-methylphenazonium methosulphate.

References

(a) DeVault & Chance (1966), DeVault *et al.* (1967); (*b*) Seibert & DeVault (1971); (*c*) Seibert (1971); (*d*) Dutton *et al.* (1971); (*e*) Hales (1976); (*f*) Kihara & McCray (1973); (*g*) Kihara & Chance (1969); (*h*) Chance *et al.* (1969); (*i*) Kihara & Dutton (1970); (*j*) Parson (1967). The rate given is for 100 μM PMS and is proportional to PMS concentration. (*k*) Floyd *et al.* (1971). See text for explanation that this is more nearly the rate of reversed primary reaction (Butler, 1973); (*l*) Butler *et al.* (1973); (*m*) Visser *et al.* (1977) observations of P-680 interpreted as oxidation of 'D_2', which may or may not be a cytochrome. The rates given may be light-limited. (*n*) Jursinic & Govindjee (1977); (*o*) Kihara *et al.* (1969); (*p*) Whitmarsh & Cramer (1979*a*, *b*). Note that plastocyanin is a probable intermediate in this electron transfer.

(*a*) *Isotope effect*. Kihara & McCray (1973) examined the effects of H_2O and D_2O on (i) cytochrome photo-oxidation in four species of photosynthetic bacteria, (ii) cytochrome reduction in one, (iii) O_2-induced oxidation of cytochromes *b* and *c* in mitochondrial preparations, and (iv) reduction of mitochondrial cytochrome *b* and flavoprotein initiated by mixing with substrate. They found that lyophilization (freeze-drying) of the bacteria or mitochondrial fragments halted cytochrome oxidation or reduction (but not the photo-oxidation of *P* in the bacteria). Reconstruction of the lyophilized samples with a small amount of H_2O fully restored the cytochrome reactions with very nearly the same rates as before drying. Reconstitution with D_2O instead of H_2O restored the amplitude of the cytochrome reactions equally well, but in all cases the half-times were increased by a factor of $\sqrt{2}$ within experimental error, usually ± 0.08. This applies to both the temperature-dependent part and the temperature-independent parts. The energy of activation was not affected. The cytochrome re-reduction (after photo-oxidation) observed in *R. gelatinosa* was biphasic and both phases showed the same $\sqrt{2}$ isotope effect. More recently they have extended their results to the reaction of cytochrome *c* with $Fe(CN)_6{}^{3-}$ (McCray & Kihara, 1979).

(*b*) *Structural relations*. Several studies of structural relations between cytochromes and neighbours should be mentioned. Most valuable is the interpretation of distances from EPR measurements of the strength of spin–spin interactions by Tiede, Leigh & Dutton (1978) (Dutton *et al.*, 1979). They conclude that the distance between each of the four cytochrome molecules able to donate directly to the RC of *C. vinosum* and the BChl of dimer of the RC is 25 Å or more. The observation is that no interaction between electron spins on these cytochromes and electron spins on '*P*' can be detected when unpaired spins are known to be present on both. They also find evidence of close interaction between the low- and high-potential cytochromes in *C. vinosum.* We might speculate from this that the high-potential cytochrome is the only direct donor to P^+ in this species and that the apparent oxidation of low-potential cytochrome results from extremely rapid re-reduction of the high-potential cytochrome by the low-potential cytochrome. The differing behaviours of the donation to *P* at different temperatures, etc., would then be explained as the influence of the electric charge of the low-potential cytochrome on electron transfer from the high-potential cytochrome.

Itoh (1980b) estimates the positions of the cytochromes, *P*, *I*, *A* and *Q* in the membrane of *C. vinosum* from flash-induced membrane potentials monitored with merocyanin dye.

Cusanovich & Mizrahi (1979) studied electron exchanges between high-potential iron proteins ('HIPIP') on the one hand and cytochromes on the other in aqueous solutions. Various combinations of the two HIPIPs, one from *Chromatium* and one from *Rps. gelatinosa*, with cytochrome c_2 from four other species of photosynthetic bacteria and with horse-heart cytochrome *c* were tried. Second-order rate constants were measured. Three of the cytochrome c_2s gave biphasic kinetics, while the other two showed single phases only, independent of which HIPIP was used. One combination (*R. rubrum* cytochrome c_2 as acceptor with *Chromatium* HIPIP as donor) was sensitive to ionic strength, the others were not. All rate constants fell within the range 1×10^{-6} to $2.4 \times 10^{-5} \text{M}^{-1}\text{s}^{-1}$. It was concluded that the biphasic cytochromes reacted with HIPIP at at least two different sites, and that non-polar interactions were usually more prominent than polar. The iron–sulphur centres of the HIPIPs are located deep within the molecule so that their distances from the exposed haem edge of the cytochromes must have been at least 4–6 Å, and tunnelling by the electron seems likely.

(*c*) *High-pressure effects.* Effects of high hydrostatic pressure on photosynthetic systems have received preliminary attention. Chance *et al.* (1979*b*) looked at cytochrome photo-oxidation and proton uptake in *C. vinosum.* The light-induced H^+-binding by chromatophores decreased in rate and amplitude reversibly as pressure increased, shutting off completely at about 2 kbar (1 bar = 1.015 atm = 10^5 Pa). Light-induced cytochrome oxidation decreased to an extrapolated zero at 4 kbar in chromatophores, 6 kbar in whole cells and 8 kbar in a Thornber subchromatophore preparation. At the higher pressures and longer times the changes became irreversible. The rates of cytochrome oxidation were not greatly changed. The effect on oxidation of *P* was somewhat smaller and more reversible. Carotenoid shifts were also observed.

Clayton & DeVault (1972; Clayton, 1979) applied high pressures to RCs of *Rps. sphaeroides.* They found that both *P*-oxidation and fluorescence quantum efficiency were decreased indicating that the pressure increased the probability of radiationless de-excitation. The efficiency of oxidation of added cytochrome decreased 30% at 2.4 kbar and then decreased further on releasing the pressure. The same phenomenon was observed once in the experiments of Chance *et al.* with *C. vinosum* whole cells.

Kleinfeld and Feher (see comment after Clayton, 1979) have also studied the effects of pressure on *sphaeroides* RCs.

(*d*) *Miscellaneous.* Bogomolni & Klein (1975, 1979) studied the electrical properties of dried films of chloroplasts from plants and algae and of chromatophores from photosynthetic bacteria using microwave techniques. They confirm the presence and trapping of light induced charge carriers of both signs, and measure their mobilities and relative quantum yields and estimate the distance between them. The data is consistent with a picture of localized charge carriers hopping between molecules and not with a band structure. The signals are not observed in a reaction-centreless mutant of *R. sphaeroides*.

Parak *et al.* (1980) concluded from Mössbauer studies of the statistical spread of the velocities of iron atoms in chromatophores of *R. rubrum* as a function of temperature as well as of the rates of reverse electron transport from both '*A*' and '*Q*' that conformational changes in the protein are frozen below about 170 K and free-up above that temperature. This conclusion is similar to the earlier observation of Hales (1976).

The effects of the fields of surface-adsorbed ions on the potentials and the rates of oxidation of cytochromes by the RC in bacteria were studied by Matsuura *et al.* (1979, 1980) and Itoh (1980a).

Rubin *et al.* (1980) found that electric currents are induced at different temperatures on warming a sample of photosynthetic RCs in no field after it was frozen in an electric field.

4.2 Mitochondrial and other non-photosynthetic electron transport

Outside the photosynthetic systems electron transport is known to occur in the respiratory chain of redox carriers in mitochondria and in redox reactions involved in the functioning of microsomes and enzymes such as catalase and various peroxidases. These are all subcellular components. Catalase destroys peroxide. Poisonous substances are metabolized by microsomes which contain the cytochrome called 'P_{450}' and utilize oxygen and also by peroxidases which use peroxides.

Mitochondria are respiratory organelles found in the cells of all oxygen-utilizing plants and animals. They appear to have a common ancestry with aerobic bacteria that formed a symbiotic relationship with the eukaryotic (nucleus-containing) cells which they invaded. In net-effect, they use oxygen to 'burn' fuel to CO_2 and use the energy to make ATP. This is a reversal of the carbon fixation and oxygen production of photosynthesis. It could only have developed after the photosynthetic organisms had built up an appreciable supply of oxygen in the surrounding water and air. It is a much more efficient utilization of the fuel than the older, primitive, anaerobic fermentation process which converts carbohydrates into alcohol, as in yeast, or into lactic acid, as in muscle. The anaerobic fermentation yields two or three ATP molecules per glucose molecule or glycosyl unit, respectively, consumed. The aerobic 'combustion' produces at least thirty ATP molecules per glucose.

The 'combustion' is not a direct combination of oxygen with the substrate. It is done through the intermediary services of an electron transport chain, most of which looks suspiciously like the electron transport chains found in photosynthesis. The substrate feeds electrons into the chain at its low redox potential end. Cytochrome oxidase at the high-potential end passes the electrons to oxygen reducing it to water. The system is shown in Fig. 4.3. The section of the chain at the low-potential end involving NAD, Fp, Fe–S, and *Q* is very similar to the chain in the purple bacteria (Fig. 4.1) that uses ATP to drive electrons to the low potential of NADP (similar to NAD). However, in the mitochondrial case the electrons flow in the opposite direction from NAD

Fig. 4.3. The scheme of oxidative phosphorylation carried out by respiration. Electron transport is driven by the redox potential gradient from the CO_2-producing citric acid cycle to acceptance by oxygen to form water. From DeVault (1979), p. 309.

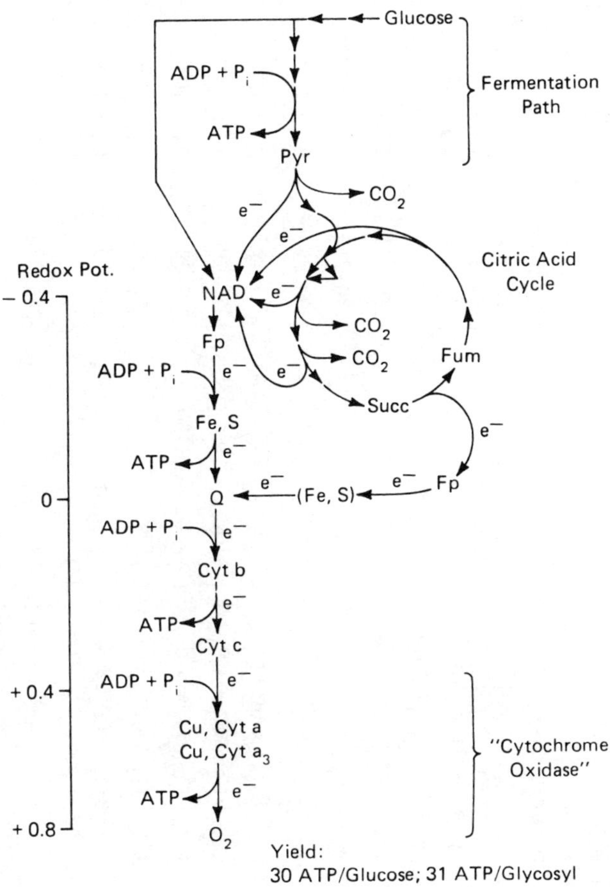

Table 4.3 *Non-photosynthetic biological electron transport*

Reactions	Range (K)	k_∞ (s^{-1})	$E^\ddagger$ (eV)	$t_{\frac{1}{2}}$ (298 K)	References
First order					
cyt. $a_3 \xrightarrow{e^-} O_2$ (Mito)	243–298	3.0×10^5	0.17	1.7 ms	(*a*: fig. 7)
cyt. $a \xrightarrow{e^-}$ cyt. a_3 (Mito)	291–312	1.3×10^5	0.16	3.2 ms	(*a*: fig. 4)
	243–291	1.0×10^{14}	0.68	(1.9 ms)	
cyt. $b \xrightarrow{e^-}$ cyt. c_1 (Mito)		$< 6 \times 10^3$	< 0.22	0.57 s	(*b*: p. 305)
succ. $\xrightarrow{e^-} F_p$; cyt. *b* (Mito)	278–333	7.2×10^7	0.50	2.6 s	(*b*: p. 306)
$A_1 \longrightarrow B_2$	170–210	4.6×10^{13}	0.54	(22 μs)	(*c*, *d*)
$A_2 \longrightarrow C_2$	170–220	8.5×10^{12}	0.50	(22 μs)	(*c*, *d*)
$Pd^- \cdot P_{450}{}^{II} \cdot Cam \cdot O_2 \longrightarrow Pd^0 + P_{450}{}^{III} + CamOH$	233–263	6.3×10^8	0.50	(0.30 s)	(*e*: p. 2838)
$P_{450}{}^{II} \cdot Cam \cdot O_2 \longrightarrow P_{450}{}^{III} + \cdots$(Aq)	273–298	1.1×10^{11}	0.78	96 s	(*f*: p. 225)
$P_{450}{}^{II} \cdot Cam \cdot O_2 \longrightarrow P_{450}{}^{III} + \cdots$(Aq + EgOH)	263–298	6.2×10^{11}	0.82	92 s	
$P_{450}{}^{II} \cdot O_2 \longrightarrow P_{450}{}^{III} + \cdots$(Aq)	275–298	2.0×10^{12}	0.82	32 s	
$P_{450}{}^{II} \cdot O_2 \longrightarrow P_{450}{}^{III} + \cdots$(Aq + EgOH)	252–298	6.3×10^8	0.63	50 s	
Second order		($M^{-1} s^{-1}$)			
$O_2{}^- \xrightarrow{e^-}$ cyt. *c*, pH < 7.45	286–320	7.3×10^9	0.22	0.5 μs·M/C	(*g*)
$O_2{}^- \xrightarrow{e^-}$ cyt. *c*, 7.45 < pH < 9.2		9.2×10^8	0.21	2.3 μs·M/C	

$P_{450}{}^{III}\cdot Cam + Pd^- \longrightarrow P_{450}{}^{II}\cdot Cam + Pd$	233–263	7.2×10^{11}	0.37	(2.0 μs·M/C)	(*e*: p. 2836)
$HRP\cdot H_2O_2 + AH \longrightarrow HRP\cdot OH + A$ (AH = luminol. In DMF + Aq.)	223–243	2.7×10^{13}	0.50	(8.0 μs·M/C)	(*f*: p. 188)
$DCPIP^0 + AH_2 \longrightarrow A + DCPIP^R$ (AH_2 = ascorbate)	238–293	2.4×10^9	0.42	4.2 ms·M/C	(*f*: p. 182)
$Ru(NH_3)_6{}^{2+} \xrightarrow{e^-}$ cyt. *c* (pH = 7, μ = 0.1)		4.7×10^6	0.12	18 μs·M/C	(*h*)
$Fe(EDTA)^{2+} \xrightarrow{e^-}$ cyt. *c* (pH = 7, μ = 0.1)	282–308	6.4×10^8	0.6	27 μs·M/C	(*i*)

Abbreviations

'e^-' over an arrow means electron transfer between the two components indicated. A_1 = cytochrome oxidase – O_2 complex in the state: $(Cu^R a^R Cu^R a_3{}^R)\cdot O_2$. A_2 = the same in the state: $(Cu^o a^o Cu^R a_3{}^R)\cdot O_2$. B_2 = the same in the state: $(Cu^R a^R Cu^o a_3)\cdot O_2^=$. C_2 = the same in the state: $(Cu^o a^o Cu^o a_3{}^R)\cdot O_2{}^-$. Superscript R means reduced, o means oxidized. Pd^-, Pd^o = reduced, oxidized putidaredoxin. Cam = camphor. P_{450} = bacterial cytochrome P_{450}. Aq = aqueous medium. EgOH = ethylene glycol. DMF = dimethylformamide. Succ = succinate. Fp = flavoprotein. Mito = mitochondria. HRP = horseradish peroxidase. DCPIP = 2,6-dichlorophenolindophenol. Cyt = cytochrome.

( ) means that a $t_{\frac{1}{2}}$ (298 K) value is extrapolated beyond the region of measurement or of validity.

References

(*a*) Erecinska & Chance (1972); (*b*) Kihara & McCray (1973); (*c*) Chance *et al.* (1975); (*d*) Chance *et al.* (1979*c*); (*e*) Hui Bon Hoa *et al.* (1978); (*f*) Douzou (1977): the P_{450} reactions are auto-oxidations in the absence of Pd. $HRP\cdot H_2O_2$ is also called 'complex I' and HRP·OH is 'complex II'; (*g*) Butler *et al.* (1975). Pulse radiolysis is the source of $O_2{}^-$ in the *in vitro* study; (*h*) Ewall & Bennett (1974) (pH = 7); (*i*) Hodges *et al.* (1974).

Note: Under $t_{\frac{1}{2}}$ for the second-order reactions, '/C' means divided by the concentration of one of the reactants to change to pseudo-first order.

to Q and make ATP instead of using it. The section from Q through cytochrome *b* to cytochrome *c* is similar to the cyclic electron-transport system of the bacteria. The chain of reactions (not electron transport) from glucose to pyruvate is very similar to the fermentation pathway mentioned above.

The citric acid cycle (Krebs & Johnson, 1937) reverses the effect of the CO_2-fixation cycle (Bassham *et al.*, 1954; Calvin, 1956) but the components are not very similar.

Data on the temperature dependence of the rate of electron transport reactions in mitochondria and other non-photosynthetic systems are collected in Table 4.3. There are no columns for a low-temperature phase as all of the reactions stop at the low temperatures. In the photosynthetic systems only the photo-reactions and dark subsequent reactions associated with the reaction centres go below 100 K and of course these are absent in the non-photosynthetic systems. Otherwise the table is constructed like Tables 4.1 and 4.2.

Several of the reactions in Table 4.3 are seen to have $k_\infty \approx 10^{13}\ \mathrm{s}^{-1}$ and some of the second order are close to $10^{11}\mathrm{M}^{-1}\mathrm{s}^{-1}$. This indicates that in them the electron transfer after the necessary nuclear rearrangement to achieve the activated state is too fast to be rate limiting. These might or might not be electron tunnelling but it does not matter because it is fast enough to keep the process adiabatic. In other cases there is indication of rate limitation by the electronic process itself, meaning that the necessity for the electron to tunnel is slowing things. As warned before, this 'indication' could be invalidated by a large negative entropy of activation or by the presence of nuclear tunnelling.

The reaction in Table 4.3 between $P_{450}{}^{III}\cdot$Cam and Pd^- (Hui Bon Hoa *et al.*, 1978, ref. (*e*), p. 2836) illustrates another pitfall. As given there it looks like a good electron-transfer reaction. However, the analysis by the authors shows that the rate is determined by a protein conformational change associated with an electron spin change (from low to high) following binding of the Pd^-. After the spin change the electron transfer is fast. Thus the reaction rate parameters given in the table for this reaction do not apply to electron transfer, but this would not be known without the more detailed analysis. In another sense it illustrates how electron-transfer rates are usually controlled by nuclear movements, this case being more complicated than just waiting for enough thermal activation of the vibrations.

4.2.1 *Interfaces between proteins*

The interfaces between reacting proteins have been investigated by several groups. One would hope to learn how close the redox centres of the molecules are when in the reacting configuration and also how the specificity of donor for acceptor is achieved. Salemme (1976, 1977, 1979) using X-ray determined structures of cytochromes and non-haem iron (iron–sulphur) proteins, notes how the protein-wrapping around the redox centre couples particular vibrations to the centre and can 'tune' the effective redox potentials and electron-transfer behaviour. Most have their prosthetic groups nearly exposed on one side, an exception being 'HIPIP' a high-potential iron–sulphur protein in which the redox centre is deeply buried. However, when one fills in all the polypeptide side-chains the exposures are not so great that direct contact between redox centres on neighbouring proteins can be assumed. He studied the closest approach possible for cytochromes *c* and b_5. These are not natural partners but they react well *in vitro*. The closest distance between π-bonded haem atoms is about 8.4 Å. Different cytochromes all tend to have a similar distribution of positively charged groups on their surface close to the haem crevice. One supposes, therefore, that these are important in determining approach to and specificity for reacting partners.

Such features were investigated by Ferguson–Miller, Brautigan & Margoliash (1978, 1979). They substituted a rather large negatively charged group, carboxydinitrophenyl, on various lysine groups on the surface of cytochrome *c*. When substituted on any lysine group in an area close to one end of the haem crevice the rate of reaction with cytochrome oxidase in mitochondria was greatly diminished.

4.2.2 *Reactions of proteins with inorganic reagents*

Feinberg *et al.* (1977, 1979) studied the effects of ionic strength on the rates of reduction of two different *c*-type cytochromes by the small inorganic reductant ferrous-$EDTA^{2+}$. From these measurements they could calculate the effective charge on the protein and compare it with the overall charge of the protein. The effective and the overall charges disagreed by only 1 unit out of 9 in one case and 1 out of 3 in the other. They concluded that the rates of reaction were determined by the overall charge of the protein and not by any localized charge at a particular reaction site. They applied the same analysis to data of Jacks *et al.* (1974) on the reduction of a rubredoxin by V^{2+} and came to the same conclusion although here the difference was nearly 2 units of charge out of 5 and one could wonder whether the difference was actually significant.

Wherland & Gray (1976) studied the reduction of cytochrome *c* (horse heart) by the 'reagents' $Fe(EDTA)^{2-}$ and by $Ru(NH_3)_6{}^{2+}$ and its oxidation by the 'reagents' $Co(phen)_3{}^{3+}$ and by $Fe(CN)_6{}^{3-}$ (EDTA = ethylenediaminetetracetate; phen = 1, 10 phenanthroline). Using the measured self-exchange rate constant for the reagents, the rate constant for the electron transfer to or from cyt *c* and the equilibrium constants from the differences in their redox midpoint potentials they calculated, using equations (3.9) and (3.10), a self-exchange rate constant, k_{11}, for cyt *c*. This was corrected by adding in the work terms calculated with various assumptions about the effective charge and size of the protein as it meets the reagent. The k_{11} values depended strongly on the reagent in the order: $Fe(CN)_6{}^{3-} > Co(phen)_3{}^{3+} > Ru(NH_3)_6{}^{2+} > Fe(EDTA)^{2-}$, encompassing a factor of 1000 from first to last. According to Marcus' theory, if all the reagents attacked the protein at the same position and distance from the haem as does cyt *c* itself, they should all yield the same true k_{11} for cyt *c*. The differences are attributed to the two fastest reagents penetrating the haem crevice by means of their hydrophobic coordination shells with $Fe(CN)_6{}^{3-}$ having the additional advantage of cylindrically orientated π-orbitals on the ligands, while $Fe(EDTA)^{2-}$ has the disadvantage that its more hydrophobic side, if turned toward the haem crevice, would put its iron atom further from the protein.

Mauk, Scott & Gray (1980) applied the same sort of analysis to the reaction of $Fe(EDTA)^{2-}$ with six different redox proteins. Assuming that $Fe(EDTA)^{2-}$ does not penetrate below the surface of the proteins they use the k_{11} values found for the different proteins to estimate the distance of their redox centres below their surfaces. To estimate distances they set k_{11} proportional to the Gamow factor, equation (2.1). They evaluate the pre-exponential factor empirically from the k_{11} value obtained for stellacyanin. They assumed that the distance between the redox centres of two stellacyanins in contact is 3.7 Å and, following Hopfield (1974), that the barrier height, V, is 2eV. The resulting relation between the depth, R_p, of redox centre below the surface of a protein and its $k_{11}{}^{\infty}$ (k_{11} corrected to infinite ionic strength) is

$$R_p = 6.2 - 0.35 \ln(k_{11}{}^{\infty}) \tag{4.2}$$

with R_p in the angstrom units and $k_{11}{}^{\infty}$ in $\mathrm{M}^{-1}\mathrm{s}^{-1}$. The depths below the surface found are: cytochrome *c* (horse heart) 5.3 Å; cytochrome c_{551} (*aeruginosa*), 5.9 Å; (stellacyanin, 1.9 Å); plastocyanin (*vulgaris*), 4.4 Å; azurin (*aeruginosa*), 7.4 Å; HIPIP (*vinosum*), 7.7 Å.

Wherland & Pecht (1978) have continued this work into studies of direct electron transfer from protein to protein, including three *c*-type cytochromes, three azurins and two plastocyanins. They find that six apparent values for k_{11} for these proteins (two azurins have the same value and so do the two plastocyanins) will excellently predict the observed rates of thirteen electron-transfer reactions between them, according to simple Marcus theory. They interpret this to indicate no special kinetic selectivity in their reactions with each other.

4.2.3 *Structural relations*

Salerno & Ohnishi (1979; Salerno *et al.* 1977) studied the spin–spin interactions in the succinate dehydrogenase portion of the mitochondrial system. They suggest the following pathway for electron transfer from succinate to quinone:

$$\text{Succ} \xrightarrow{2e^-} \text{FAD} \xrightarrow{e^-} \text{S–1} \xrightarrow{e^-} \text{S–2} \xrightarrow{e^-} \text{S–3} \xrightarrow{e^-} \text{Q} \xrightarrow{e^-} \text{Q},$$

where FAD = flavin adenine dinucleotide, S = iron–sulphur centre, and Q = quinone. The S–2 to S–3 distance could not be determined but the others appear to be 8–10 Å between nearest neighbours in the indicated chain. The slow step in the chain is Succ → FAD (about 5 ms at 18°C). A review of the data and its interpretation concerning the electron carriers in the section of the mitochondrial chain (see Fig. 4.3) between NADH, succinate–fumarate and Q has been written by Ohnishi (1979).

Gupta & Yonetani (1973) in a study of the cytochrome *c* cytochrome–*c* peroxidase reaction complex, using NMR, determined that the two haems are separated by more than 25 Å.

Erecinska and co-workers (1977, 1978, 1979) have probed the geometrical arrangement of the electron carriers in the quinone to O_2 region of the electron-transport chain of mitochondria using orientated multilayers of mitochondrial membranes and also photoaffinity labelling. By observing dichroic ratios for the absorption of polarized light at different angles to the membrane they conclude that the haem planes of cytochromes a and a_3 are perpendicular to the membrane planes. This overrules a tentative opposite conclusion on the basis of laser-induced dichroism by Junge & DeVault (1975). The photoaffinity labels showed which subunits of the cytochrome oxidase meet cytochrome *c* when the latter passes electrons to the former. Vanderkooi & Erecinska (1975) and Vanderkooi & Landsberg (1977), studying energy transfer by Förster mechanism from fluorescent modifications of cytochrome *c* to cytochrome oxidase, conclude that the distance between the haems of cytochromes *c* and a, a_3 is 15–20 Å. Van Gelder & Beinert (1969) concluded from EPR measurements that the distances between the haems of cytochromes a and a_3 and their associated Cu atoms are small, while Leigh & Wilson (1972) and Leigh *et al.* (1974) suggested on the basis of photodissociation effects of the CO compound that the haems of a and a_3 may be parallel and share a common pocket. Other EPR measurements probing the spatial relations between the redox carriers in cytochrome *c* oxidase are Van Gelder *et al.* (1967), Wilson, Frecinska & Owen (1975) and Wever, Van Gelder & Der Vartanian (1975).

Brill (1978, 1979) analysed the EPR spectrum of the Cu-containing protein azurin. He fits the data with a Gaussian distribution of the angles of the ligands to the copper, and from this he estimates a force constant. This is a parameter that can be used to estimate a value for λ (see equation (3.4)) which would be involved in electron transfer to and from the copper atom.

4.2.4 *Miscellaneous electron transfers*

Pulse radiolysis has been used extensively to study the reduction of ferricytochrome *c*. In oxygen-free aqueous solutions, pulses of high-energy electrons form hydrated electrons which reduce the cytochrome with rate constants of 1 to $2 \times 10^{10}\ \text{M}^{-1}\text{s}^{-1}$ at pH 13 to 6 (Land & Swallow, 1971). In oxygen-saturated solution, the superoxide anion radical, $O_2^{\,-}$, is formed and reduces cytochrome *c* with $k = 1.4 \times 10^6\ \text{M}^{-1}\ \text{s}^{-1}$ for the protonated form that is predominant from pH 4.7–7.45, with $k = 3.0 \times 10^5\ \text{M}^{-1}\ \text{s}^{-1}$ for the form predominant between pH 7.45 and 9.2 and not at all with the form prevailing above pH 9.2 (Butler *et al.*, 1975; ref. (*g*) in Table 4.3). This study was done in further detail by Koppenol *et al.* (1976), who rejected electron tunnelling on the basis that the reaction has a noticeable temperature coefficient, which, however, is not a valid reason.

Experiments on electron transfer to and from artificial lipid vesicles and micelles may throw light on the tunnelling process since they provide opportunity to vary the parameters. The work of Henglein and associates already mentioned in Chapter 3, is leading into studies of biological types of molecules and models for the photosynthetic system (Alkaitis, Grätzel & Henglein, 1975). Vanderkooi (1976) has made similar studies in which she photolyses a dye dissolved in lipid vesicles and observes the subsequent oxidation of cytochrome *c* in the aqueous surroundings. The supposition is that the electron tunnels from the cytochrome through the lipid–water interface to the photoionized dye molecule.

Kowalski (1965), using nuclear magnetic resonance, measured the rate of exchange between ferri- and ferrocytochrome *c* in 0.02 M solution as $100\ \text{s}^{-1}$.

Ballard & Mauzerall (1979, 1980) studied the photolysis of zinc octaethylporphyrin (ZnOEP) in solvents of various viscosities and dielectric constants. With the right conditions they can observe the formation of triplet states following absorption of a 5–7 ns pulse of laser light. By following the electrical conductivity of the solution they observe the formation of positive and negative ions of ZnOEP by electron transfer between pairs of triplet-excited ZnOEP molecules. Analysis of the data in three independent ways gives a reaction radius for the electron transfer of 21 Å. The physical encounter radius of the ions is 15 Å. The conclusion is that the electrons tunnel through up to 7 Å of intervening solvent.

EPR signals have been associated with cancerous tissue. Floyd, Bronsdon & Commoner (1973) observed an association with ascorbate free-radical. Swartz & Dodd (1981) summarize data on the ascorbyl radical and conclude it to be the principal long-lived radical produced in living systems and that its relatively non-toxic properties make ascorbic acid a good anti-oxidant. Floyd & Soong (1977) and Floyd *et al.* (1978) studied the occurrence of nitroxyl free-radical in cancerous tissue. Since the free-radicals so detected may arise from electron transfer there is a possibility that electron-transfer mechanisms could become important in the study of cancer. An old study of charge-transfer spectra of aromatic hydrocarbons (Mason, 1959; Birks, 1959) showed that the most carcinogenic of them had spectra which could be interpreted (Pullman & Pullman, 1958; Pullman, 1979) as indicating a redox midpoint potential for the hydrocarbon falling in a narrow range. The suggestion was that the cancer is caused by electron transport between the hydrocarbon and some cell constituent having this particular midpoint potential.

It is thought that cancer can be caused by removal of a protein protective coating over pieces of DNA encoding cancerous information (Busch, 1962). Laki & Ladik (1976) propose that holes (absence of electrons) which can be produced in proteins by charge-transfer acceptors will allow electronic excitations which greatly increase the electronic polarization and dispersion forces of attraction between the molecules. Ladik (1978) suggests that electron donors could initiate cancers by filling the holes and releasing the protein from the DNA.

As mentioned earlier, a study of electron semiconduction induced in proteins by methylglyoxal is reported by Pethig & Szent–Györgyi (1977).

Semiconduction in biological materials has received much attention since the suggestion of Szent–Györgyi (1941) that it should exist. Pethig (1979) has written a good review and exposition of this and other electrical properties of such materials on which I will rely for much of the following information. Actually, when placed between electrodes most biological materials are better described as insulators. Typical materials have high resistivity (proteins: 10^{17}–10^{20} ohm cm; nucleic acid salts: 10^{7}–10^{17} ohm cm) at room temperature and the activation energies are usually large (1.0–1.5 eV) (Pethig, 1979, pp. 291, 307). The conductivity increases exponentially with the amount of water absorbed. Mobility measurements sometimes suggest conduction in bands as illustrated in Fig. 2.2A, but more often indicate electron or hole hopping from one localized site to another. See Lewis (1977) and Petrov (1979) for reviews.

Charge-transfer complexes greatly increase the conductivities. They promote the initial charge separation and can act as extrinsic donors or acceptors (as in the 'doping' of silicon with impurities) to produce *n*-type or *p*-type semiconduction respectively. Water is believed to act as an electron donor to protein (Eley & Leslie, 1964, 1966). For example, the conductivity of dry haemoglobin is *p*-type, but the very small concentration of acceptors is swamped by the donor activity of even a small amount of water, converting it to *n*-type (Trukhan, 1966). Eley & Pethig (1971) observed a similar effect in DNA. At much higher concentrations of water, protonic conduction can dominate.

Methyl glyoxal combines by Schiff-base linkage to lysines and acts as a charge-transfer acceptor colouring proteins brown and greatly increasing their conductivity (Pethig & Szent–Györgyi, 1977; Otto, Suhai & Ladik, 1977; Otto *et al.*, 1978; Bone *et al.*, 1978; Pethig, 1978). The holes produced in collagen by methyl glyoxal have mobilities indicative of band structure. A dielectric dispersion observed at low frequencies is interpreted as meaning that the holes are confined to conducting regions within the protein.

It is proposed (Ladik, 1974, 1975) that a charge transfer may occur between the bases and the sugar-phosphate 'backbone' of DNA causing electronic conduction down the backbone and hole transport across the base-pairs.

Melanins, the black pigments in skin, hair and retinas, have very interesting electrical properties. The black colour, which extends far beyond the visible range, suggests efficient coupling of electronic excitation to vibrational modes (McGinness & Proctor, 1973). They contain 10^{19} unpaired electrons per cm^3. Their electrical conductivity increases markedly with application of even small electric fields (Blois, 1971). This allows a memory-type switching from resistive to conductive states at thresholds of the order of $300\ V\ cm^{-1}$. It is speculated that such behaviour could be useful in sense organs transducing various stimuli into nerve impulses (McGinness, Corry & Proctor, 1974; Filatovs, McGinness & Corry, 1976). In fact melanins are found in the midbrain and inner ear and their loss may contribute to Parkinson's disease, schizophrenia and deafness (Lindquist, 1973; Proctor, 1972; Proctor, McGinness & Corry, 1974). Melanin consumes oxygen at appreciable rates. Sarna *et al.* (1980) have studied the factors involved.

Superconductivity has been considered for aromatic compounds (London, 1937), for organic polymers (Little, 1964), and for DNA (Ladik, Biczó & Rédly, 1969). Some bile cholates showed apparently perfect diamagnetism in small domains (Halpern & Wolf, 1972; Wolf, 1976) and experiments with lysozyme (Ahmed *et al.*, 1975, 1976) suggested that a small superconductive region existed in each lysozyme molecule and that a cluster of such produced co-operative phenomena. The subject is still controversial.

There have been several suggestions without much experimental evidence that electron tunnelling may be involved in nerve action. Without themselves implying electron tunnelling in nerves, Mueller & Rudin (1963) pointed out that the (ion) current v. voltage characteristics of nerve membranes had the same shape (negative resistance, etc.) as the *I–V* plot of the tunnel-diode, Fig. 2.2 B. Wei (1967) then suggested that nerves actually used electron tunnelling. Walker (1977) made a more detailed study and suggested that electron tunnelling might initiate opening of the gates that release the transmitter substance at the synapse. He backed his theory with statistical studies of nerve-firing frequencies. The feeling of neurophysiologists is that ionic conduction can account for all the main observed electrical currents, but there are also small 'gating currents' (Armstrong, 1974) apparently connected with the opening of the sodium channels as the action potential moves along the nerve, and these are still open to various interpretations (Wei 1969, 1971; Petrov, Teslenko & Kharkyanen, 1978; Lee, 1980).

4.3 Charge-transfer bands

Charge-transfer or intervalence absorption bands were briefly described in chapter 3. See also 'CT' in Fig. 3.1, and explanation of the theory in the appendix. Observation of such should provide independent measurement of electron-transfer parameters. The intensity of a charge-transfer band is proportional to the square of the Hamiltonian matrix element for the electron transfer and so is expected to be very weak for highly non-adiabatic electron transfers such as many of those encountered in biological systems. Direct measurement of light absorption would probably be impossible.

Hopfield (1976, 1977, 1979) proposed to increase the sensitivity of the measurement by looking for much larger changes in the absorption spectra of donor and acceptor which would accompany the transfer of an electron back from acceptor to donor when illuminated in the charge-transfer band. The sensitivity was further enhanced by chopping the illumination and use of a lock-in amplifier.

By this means Potasek & Hopfield (1977*a*, *b*) and Potasek (1979) carried out the measurements on two systems, namely cytochrome *c*–iron hexacyanide and cytochrome *c* peroxidase–cytochrome *c* and at first reported positive results. However, Austin & Hopfield (1982) repeated the experiment with cytochrome *c*. They illuminated with 15 ns pulses from a Nd:YAG laser (1064 nm wavelength) to increase the sensitivity further and found no effect of the expected charge transfer which would have been prompt. They did find a large absorption at 1.06 μm which appeared when they mixed their $Fe(CN)_6{}^{4-}$ and $Fe(CN)_6{}^{3-}$. This absorption of light heated the reaction mixture and caused a non-prompt (~ 1 ms) shift in the equilibrium between the oxidation states of the cytochrome *c* and those of the iron hexacyanides. This could be mistaken for photo-induced charge transfer when not sufficiently time resolved as in the experiments using chopping. They suggest also that photo-reduction by TRIS buffer may have interfered with the peroxidase experiment.

Chang & Austin (1982) have extended the cytochrome *c*–FeCy work and find a charge transfer induced by light absorbed into the visible haem bands (from an excited electronic state), but still only the heat-induced response in the infra-red. They obtained similar results with ruthenium–cytochrome *c* covalently bound complex. After further examination of the theory they conclude that it predicts weaker charge-transfer bands than they would be able to detect.

4.4 Biological control

Chance (1979) emphasizes the importance of control in biological systems. An illuminating discussion of the general subject is the book *Cybernetics* by Wiener (1948). The point is that all biological processes must be controlled so that one will not get ahead of the others and cause chaos. In particular, electron transport in the mitochondrial chain must stop when the ADP is all phosphorylated or the system will burn up. Suggestions have been made that equilibrium between electron transport and phosphorylation tightly coupled to it is sufficient (Slater, 1953; Klingenberg & Schollmeyer, 1963) and this has strong support for application to the section of the mitochondrial chain between NAD and cytochrome *c*, where it was shown by Chance & Hollunger (1957, 1961) that added ATP can reverse the flow of electrons driving them to NAD. However, in the cytochrome oxidase region the redox potential of O_2/H_2O is so high that control by equilibrium with ATP/ADP seems unlikely and a kinetic control is favoured. The unanswered question is whether knowledge of the mechanism of electron transfer can help in understanding biological control of the process.

In this connexion, DeVault *et al.* (1967) suggested that the requirement for energy matching in electron tunnelling could furnish a means of kinetic control. See Blumenfeld & Chernavskii (1973) for a similar suggestion. The supposition is that electron transfer between a donor and an acceptor could be turned on or off by application of an electric field that altered the electrical potential of donor with respect to acceptor to cause matching or mismatching of the pertinent energy levels. The electric field could come from a membrane potential or from movement of ions or polar groups to or from the redox centres,[†] and for control purposes would be part of a feedback loop. When, however, one considers the vibronic interactions which allow energy matching over a range of thermal fluctuations it is seen that such control cannot be as sharp as the quantum-mechanical requirements for energy matching ('resonance') in a static system. Instead, one has the 'energy-gap' law which, in its simplest form is given by equation (3.3). Changing the relative electrical potential between donor and acceptor is an alteration of ΔG° in the equation. The effect on rate is found by substituting equation (3.3) into (3.1). As already discussed in Chapter 3 the energy-gap law often tends to fail when $-\Delta G^\circ$ is greater than λ. However, it seems to hold reasonably well for $-\Delta G^\circ < \lambda$ and this will be sufficient for present considerations.

Because we are interested in *kinetic* control we consider only reactions far enough from equilibrium that the reverse rate can be neglected. We define the control as: $C = \mathrm{d}(\ln k)/\mathrm{d}(\Delta E/k_\mathrm{B}T)$. We use $\ln k$ because we are interested in fractional or relative changes in rate. The symbol ΔE is equivalent to and used in place of $-\Delta G^\circ$ to emphasize its dependence on electrical fields. It is divided by $k_\mathrm{B}T$ to measure it in units of $k_\mathrm{B}T$. This definition of C is seen to be equivalent to the Brönsted (1928) coefficient of catalysis, and also to be related to the Tafel (1905) coefficient in electrolysis. Differentiating the expression for $\ln k$ obtained from equations (3.1) and (3.3) with respect to $(\Delta E/k_\mathrm{B}T)$ gives:

$$C = \frac{\mathrm{d}(\ln k)}{\mathrm{d}(\Delta E/k_\mathrm{B}T)} = \frac{1}{2}\left(1 - \frac{\Delta E}{\lambda}\right). \tag{4.3}$$

For $\Delta E \sim 0$, $C \approx \frac{1}{2}$ and diminishes to 0 at $\Delta E = \lambda$. This means that ΔE must be shifted by at least 0.12 eV at room temperature to change the rate by a factor of 10, or by 0.24 eV for a factor of 100. Local fields intense enough to produce this much potential difference between donor and acceptor 1 or 2 nm apart might be possible. (A single electronic charge

[†] In this connexion the studies of Itoh (1978) (Matsuura *et al.*, 1979) on the effects of double-layers and salt concentrations at the surface of membranes on rates of electron transport to and from the membrane may be pertinent.

in a vacuum produces a gradient of 0.12 V/nm at a distance of 3.5 nm.) The possibility of enhancing the control factor, based upon theoretical considerations of Sarai (1979, 1980) will be discussed in Chapter 5. (See Section 5.4 and Fig. 5.6).

Aside from purposeful control in feedback loops, we should also consider the possibly casual control of electron-transfer rates that occurs when charges on neighbours of the donor or acceptor undergo a change. Whitmarsh & Cramer (1979*a*, *b*) have applied such consideration to the problem of the rate of cytochrome *f* re-reduction following oxidation by P_{700} in PS-I of chloroplasts. Omitting cytochrome *b*, suggested in Fig. 4.2, they take their system as:

$$Q \xrightarrow{e^-} f \xrightarrow{e^-} \mathrm{PC} \xrightarrow{e^-} \mathrm{P}_{700},$$

where Q stands for plastoquinone, a pool of 8–12 equivalents for each of the other components, and PC stands for plastocyanin. Starting with Q in largely reduced state and the other three components oxidized (after a short period of illumination) the light is shut off and three electrons transfer from Q to reduce the other three components. The kinetics of the cytochrome *f* reduction are best fitted by assuming rate constants for the transfer from Q to f which are different for each of the three electrons in the ratio of 3:2:1 from first to last. Since the charges on PC and P_{700} are becoming less positive (or more negative) during the process they should exert an electrostatic field which will be stronger on f than on Q (assuming Q is geometrically further from PC and P_{700} than f) and will shift the apparent redox midpoint potential of f downwards with respect to that of Q. This makes the energy gap, ΔE, for the $Q \rightarrow f$ electron transfer smaller for the successive transfers. Using equations (3.1) and (3.3) with $\lambda = 2.5$ eV and successive ΔEs of 0.40, 0.375 and 0.335 eV they obtain the desired ratio of rate constants.

It was suggested earlier that all cytochrome photo-oxidation in *C. vinosum* might go through the high-potential cytochrome *c*. If so, the rate of the transfer, $C(\mathrm{h.p.}) \xrightarrow{e^-} P^+$, would depend upon the charge state of the low-potential cytochrome in agreement with the observed half-times of 1 μs with C(l.p.) initially reduced and 2 μs oxidized. However, there are complications in the temperature dependencies which may not be fitted by a simple theory.

4.5 Energy transduction

Several authors have suggested that electron tunnelling might furnish a way to transduce the energy to a form that could be used, say, for phosphorylation. Cartling & Ehrenberg (1979) suggests that the relaxation following electron transfer at the activated configuration (*C* in Fig. 3.1) might be partly through conformational states of the proteins and only partly through the vibrational states ordinarily postulated, and that the conformational energy could be stored (delayed relaxation) and so utilized for ion transport or phosphorylation, etc.

Chernavskaya *et al.* (1977) also suggested that mechanical transduction of energy to cause phosphorylation can come from reception of an electron in a site, T_R, followed by relaxation of the environment of the site, i.e. polaron formation. This electron movement is confined to a two-site system contained within the ATPase and isolated from the electron-transport chain. It is conceived of as influenced, however, by Coulomb forces from electrons in the chain. The electron in the ATPase is pushed from T_L to T_R by such forces when a nearby carrier in the chain is occupied by an electron. When the electron moves on in the chain the electron in the ATPase moves back to T_L, and the polarization of site T_R relaxes to get ready for another cycle.

Chernavskaya & Chernavskii (1978) have also suggested that a similar local electron movement may be involved in the energy transduction in the purple membrane of the salt bacteria, *halobium*. However, this is contrary to the thinking of most people in the field.[†]

Fröhlich (1968) has suggested that biological molecules might store energy in certain longitudinal vibrational modes of low frequency (3–30 cm^{-1}) that would not readily thermalize with the other vibrational modes. Davidov (1982) suggests that solitons might do this.

Cope & Straub (1969) suggested that phonons might carry energy from electron transport to phosphorylation. However, the possible mode of action of such a mechanism is not explained and the work suffers from confusions such as attribution of the data of DeVault & Chance (1966) to 'cytochrome oxidase'.

A review of mechanisms of energy transduction in general has been given by Blumenfeld (1978).

4.6 Nuclear tunnelling in biology

We have already mentioned that nuclear tunnelling may accompany electron tunnelling, especially in the temperature-independent

[†] For example, Honig *et al.* (1979) on the basis of considerable research in the field, offer the theory that the retinal in either rhodopsin or bacteriorhodopsin is held more or less firmly at the end containing the ring and that the other end, tied by a protonated Schiff base to the flexible side-chain of a lysine residue, whips around when absorption of a photon causes isomerization from *cis* to *trans* in rhodopsin and *trans* to *cis* in bacteriorhodopsin. The motion of this ionic charge involves breaking an electrostatic bond with a fixed counter ion, in other words, a separation of charge, and can cause movement of other charged particles including protons.

low-temperature region of an electron-transfer process which shows temperature dependence at higher temperature. Thus nuclear tunnelling without electron transport is important not only for itself but also for the elucidation of the similar process in electron transport.

The polymerization of formaldehyde by nuclear tunnelling discovered by Goldanskii, Franck-Kamenetskii & Barkalov (1973) (Goldanskii, 1979*a*) was mentioned above in Chapter 3, together with implications that it has for biology.

Proton tunnelling may be quite common, but a particular role for it in the production of biological mutations has been postulated by Löwdin (1963), who proposed that these are caused by tunnelling of the protons of the hydrogen bonds joining a nucleotide base on one strand of DNA molecule to its mate on the paired strand. This gives a tautomeric form of the base which is supposed to result in mispairing of bases if it happens at the moment of replication. This mechanism has been given a new, more dynamic, quantum-mechanical treatment by Dogonadze, Kharkats & Ulstrup (1976).

Frauenfelder (1979) and co-workers (Austin *et al.*, 1974, 1975; Alberding *et al.*, 1976) are studying the binding of ligands (CO, O_2) to haemproteins (myoglobin, haemoglobin (β-chain), carboxymethylated cytochrome *c*). They first dissociate the ligand with a laser flash and then observe the course of the recombination with a spectrophotometer. They have equipped their apparatus to read-out in a logarithmic time scale in order to encompass in one sweep a time range from 2 μs to 1 ks – nearly 9 orders of magnitude.

The recombination goes by several steps at room temperature but below about 160 K there is usually only one step. In most cases the progress of the reaction follows roughly a power law:

$$C(t) = C(0)(1 + t/t_0)^{-n}, \tag{4.4}$$

where $C(t)$ is the concentration of unreacted molecules at time t, and $C(0)$ the concentration immediately after the flash at time $t = 0$. This gives a nearly linear plot when log C is plotted against log t. These slopes decrease with decrease of temperature until at about 20 K and below there is almost no dependence on temperature. This is interpreted as tunnelling of the ligand or, better, of the system, through the final barrier to combining with the iron atom. This work is especially interesting because of the detailed methods of data analysis which the authors have worked out and for the large amount of information they have extracted from the data. A good review of the techniques is given by Frauenfelder (1978).

The logarithmic time-course is explained as the result of an ensemble of first-order reactions with a distribution of values for the rate constants. This, in turn, is interpreted as a distribution of barrier heights. It is supposed that each of the protein molecules is frozen in a different conformation state giving a different barrier height. Between about 30 and 160 K the vibrational energy is sufficient that reaction is over the barrier, following Arrhenius law. In this region the distribution of activation energies can be obtained by inverting the empirical rate law with a Laplace transform.

Below 20 K the barriers are tunnelled and the distribution of barrier heights found in the higher temperature region can be used to predict the tunnelling behaviour. The circumstance that a distribution of barrier heights is available allows separation of the pre-exponential and exponential parts of the tunnelling expression. Frauenfelder (1979) also suggests that the vibrations involved may be plate-like vibrations of the haem.

The reason that distribution of parameters is not observed at higher temperature (above 160 K) is that the proteins become more mobile, the conformations are not frozen and each molecule displays a behaviour that is an average of all the conformations and therefore more uniform.

Frauenfelder *et al.* (1979) have interpreted X-ray diffraction data of protein crystals in terms of the variation in the positions of the atoms and the dependence of this variation on temperature. Maps are presented of the variabilities of the main chains and side-chains of each residue in myoglobin. Conformational variations are at least partially distinguishable from vibrations. They observe a condensed core around the haem, semiliquid regions towards the outside and possible pathways for ligands.

Frauenfelder's group (Alben *et al.*, 1980) find that substituting ^{18}O for ^{16}O in the carbon monoxide slows the rate of ligand binding to myoglobin at 20 K by 16% while substituting ^{13}C for ^{12}C slows it 35%. The large isotope effects are expected for tunnelling (equation (2.1)) and it appears that the motion may be a rotation involving the carbon more than the oxygen.

Trautwein and co-workers (Marcolin, Reschke & Trautwein, 1977) have used Mössbauer techniques to study the kinetics of the myoglobin–CO reaction at low temperatures.

Jortner & Ulstrup (1979) have worked out a relationship between the Gamow tunnelling formula, equation (2.1), and the Franck–Condon overlap factors when they involve nuclear tunnelling as found in the multiphonon theories described in Chapter 5. See Section 5.6.1(*a*).

5

Theory pertinent to biological electron transfer

Considerable basic theory has already been introduced in the previous chapters. It was necessary to explain the significance of the data presented in them. Chapter 1 mentioned that we consider a tunnelling barrier to be any classically forbidden arrangement through which it is necessary for a system to pass in order to get from its initial state to a final state. It is not necessarily a permanent structure through which a particle passes, although it can be. Fig. 2.1 is a diagram showing tunnelling of a fixed barrier and equation (2.1) gives the probability. Equation (2.2) gives a more general formulation, the 'golden rule', for rate of transition from a state with one electronic configuration to a state with another electronic configuration. The transition might or might not be by tunnelling. The key quantity in this expression is the Hamiltonian matrix element, H_{ab}, for transition from state a to state b.

The Franck–Condon principle (fast electronic movement, slower nuclear motion) and non-equilibrium polarizations produced by sudden electron transfer before the nuclei can adjust were mentioned in Chapter 2. These are further discussed in Chapter 3, along with the concept of adiabaticity with the use of Fig. 3.1, which is a diagram illustrating the nuclear motions necessarily accompanying electron tunnelling. I will divide the development of theory into three stages: 'classical', 'semi-classical' and 'quantum-mechanical' according to the treatment of the nuclear motions.

5.1 Classical

As already mentioned, this approach was developed in the late 1950s by R. A. Marcus (1956, 1965, 1979*b*). He considered a multi-coordinate model with a variety of frequencies and applied a full statistical mechanical treatment. However, with the use of a simplified one-dimensional model we can arrive at equations of the same final form.

First, a simplified derivation of equation (3.3). Fig. 5.1 is a suitable redrawing of Fig. 3.1. In Fig. 5.1 we have taken the R and P curves to be equal parabolas corresponding to harmonic oscillators with the same frequency, and we do not here consider changes of frequency.

The equations of the curves are:

$$V_R = k_{\mathrm{H}}(X - X_A)^2/2 + E_R \tag{5.1}$$

and

$$V_P = k_{\mathrm{H}}(X - X_B)^2/2 + E_P, \tag{5.2}$$

where V_R is the nuclear potential energy, that is, the total energy of the system, donor + acceptor + surroundings, minus the kinetic energy of the nuclei which are coupled to the electron transfer, when the electron to be transferred is in the donor. V_P is the nuclear potential energy when the electron is in the acceptor. k_{H} is the Hook's law force constant for the vibrations. E_R and E_P are the minimum values for R and P, respectively, and X_A and X_B are the respective equilibrium values of the oscillating

Fig. 5.1. Nuclear motion accompanying electron transfer. A re-drawing of Fig. 3.1. to label the quantities entering equation (3.3) and to show a simplified derivation of it.

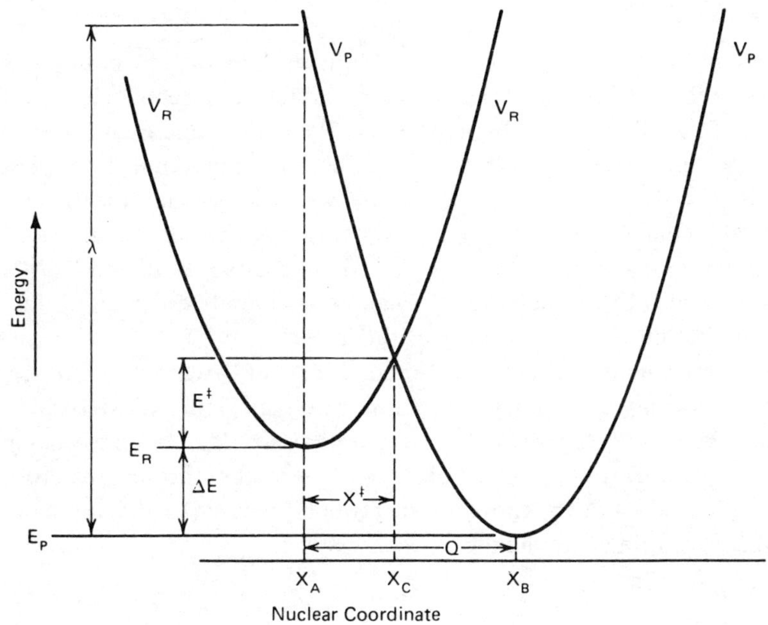

coordinate, X. The angular frequency of oscillation is given by:

$$\omega = (k_H/M)^{\frac{1}{2}}, \tag{5.3}$$

where M is the reduced mass of the oscillator.

A measure of the coupling of the nuclear vibrations to the electron transfer is the amount, $Q = X_B - X_A$, by which the equilibrium positions are shifted. As explained before, this is the effect of the Coulomb forces from the transferred electron acting on the polar groups of the oscillators. Another measure of the coupling is the energy, λ, required to displace the system an amount Q without electron transfer:

$$\lambda = k_H Q^2/2 = M\omega^2 Q^2/2. \tag{5.4}$$

Let the energy of activation be $E^\ddagger$ and the oscillator displacement from A to C be $X^\ddagger = X_C - X_A$. Then

$$E^\ddagger = k_H X^{\ddagger 2}/2. \tag{5.5}$$

Calculated as a displacement in the P system we have:

$$E^\ddagger = k_H(Q - X^\ddagger)^2/2 - \Delta E, \tag{5.6}$$

where $\Delta E = E_R - E_P$, the exothermicity of the reaction. ΔE is also called the 'energy gap'. Its sign is chosen to conform to the usual usage in electron-transfer theory, but tentatively and reluctantly because of the opposite usage in thermodynamics. Combining (5.4), (5.5) and (5.6) gives

$$E^\ddagger = \frac{(\lambda - \Delta E)^2}{4\lambda}. \tag{5.7}$$

$E^\ddagger$ in equation (5.7) is identified with $\Delta G^\ddagger$ in equation (3.3) by noting that $\Delta S^\ddagger$ is 0 if the activated state has the same vibrational frequencies as the R state. This follows since, by hypothesis, P has the same frequencies as R and the activated state is treated as a combination of R and P. Likewise $-\Delta E$ is identified with ΔG° because ΔS° is 0 when, as assumed, the frequencies do not change. In the condensed systems in which we are interested, the change of pressure–volume product is negligible. Equation (5.7) is, therefore, equivalent to equation (3.3) and this simplified derivation may give some feeling for the meaning of the latter.

Next we evaluate the factor κB in equation (3.1), treating the nuclear motion classically but using the quantum-mechanical formula derived by Landau (1932) and Zener (1932) for the probability P_{RP}, of an electronic transition from R to P upon one passage of the system through the cross-over point at X_C. For the case that the probability is small and, therefore, that the transition is non-adiabatic their result is:

$$P_{RP} = \frac{2\pi}{\hbar}\,|H_{RP}|^2\frac{1}{v\Delta F}, \tag{5.8}$$

where v is the velocity of passage (along the horizontal axis of Fig. 5.1) and

ΔF is the difference in slope of the curves R and P at X_C. This can be related to the 'golden rule', equation (2.2), by a semi-derivation given in the appendix.

If the oscillating system in the R state has total vibrational energy equal to E the classical amplitude will be $X_M = (2E/M\omega^2)^{\frac{1}{2}}$ and the velocity v, as it passes the displacement $X^\ddagger$ will be $\omega(X_M{}^2 - X^{\ddagger 2})^{\frac{1}{2}}$. Since, if $X_M > X^\ddagger$, it passes $X^\ddagger$ twice per cycle, the rate of conversion from R to P at energy E will be $W(E) = 2P_{RP}\,\omega/2\pi$ or

$$W(E) = \frac{2|H_{RP}|^2}{\hbar\Delta F(X_M{}^2 - X^{\ddagger 2})^{\frac{1}{2}}}. \tag{5.9}$$

Averaged over a Boltzmann distribution of energies, E, the mean rate, W, becomes:

$$\begin{aligned} W &= \frac{\displaystyle\int_{E^\ddagger}^{\infty} W(E)\exp(-E/k_\mathrm{B}T)\,\mathrm{d}E}{\displaystyle\int_0^{\infty}\exp(-E/k_\mathrm{B}T)\,\mathrm{d}E} \\ &= \frac{|H_{RP}|^2\omega(2M)^{\frac{1}{2}}}{\hbar\Delta F(k_\mathrm{B}T)^{\frac{1}{2}}}\int_{E^\ddagger}^{\infty}\frac{\exp(-E/k_\mathrm{B}T)}{(E - E^\ddagger)^{\frac{1}{2}}}\mathrm{d}E \\ &= \frac{|H_{RP}|^2\omega(2M)^{\frac{1}{2}}}{\hbar\Delta F(k_\mathrm{B}T)^{\frac{1}{2}}}\exp(-E^\ddagger/k_\mathrm{B}T). \end{aligned} \tag{5.10}$$

From the geometry of Fig. 5.1:

$$\Delta F = \mathrm{d}(V_R - V_P)/\mathrm{d}X = k_\mathrm{H}Q = \omega(2\lambda M)^{\frac{1}{2}}. \tag{5.11}$$

Combining (5.10) and (5.11):

$$W = \frac{|H_{RP}|^2}{\hbar}\left(\frac{\pi}{\lambda k_\mathrm{B}T}\right)^{\frac{1}{2}}\exp(-E^\ddagger/k_\mathrm{B}T). \tag{5.12}$$

Substituting (5.7) into (5.12) gives:

$$W = \frac{2\pi}{\hbar}|H_{RP}|^2(4\pi\lambda k_\mathrm{B}T)^{-\frac{1}{2}}\exp\left[-\frac{(\lambda - \Delta E)^2}{4\lambda k_\mathrm{B}T}\right]. \tag{5.13}$$

Equation (5.13) is thus equations (3.1) and (3.3), combined, with

$$\kappa B = \frac{2\pi}{\hbar}|H_{RP}|^2\frac{1}{(4\pi\lambda k_\mathrm{B}T)^{\frac{1}{2}}} \tag{5.14}$$

as given by Marcus (1979*b*). Since the vibrations have been treated classically, the result is valid only at temperatures high enough that the vibrations are fully excited.

If all of the vibrational modes in a multi-mode system are considered to have the same frequency the use of a single nuclear coordinate in

Fig. 5.1 is not actually restrictive. The coordinate can be a composite coordinate involving all of the modes and λ would be a sum of the individual λs. Further discussion of the nature of λ was given earlier. Marcus' derivation of equations (3.1) and (3.3) with elaborations for special cases, involved setting up a statistical mechanical model of the transition state, itself. It should be consulted (Marcus, 1965) for more exact definitions of the parameters in terms of multi-dimensional, multi-oscillator systems with both frequency and equilibrium position changes.

5.2 Semiclassical

To explain temperature independence at low temperature while temperature dependence occurs at high temperature it is necessary to treat the oscillators quantum mechanically. For example, a set of oscillators in thermal equilibrium calculated by classical mechanics will show average energy equal to $k_B T$ and probability of having displacement, x, from equilibrium equal to:

$$P(x) = (2\pi\sigma_x{}^2)^{-\frac{1}{2}} \exp\left(-\frac{x^2}{2\sigma_x{}^2}\right), \tag{5.15}$$

which is a Gaussian distribution with the square of standard deviation equal to:

$$\sigma_x{}^2 = k_B T / k_H. \tag{5.16}$$

The quantum-mechanical treatment takes account of the fact that the energy levels are quantized with a spacing between levels equal to $\hbar\omega$ and with the lowest level at $\hbar\omega/2$. It gives for the average energy:

$$\bar{E} = (\hbar\omega/2)\coth(\hbar\omega/2k_B T) \tag{5.17}$$

and for the probability of displacement, x, the same as equation (5.15) but with:

$$\sigma_x{}^2 = (\hbar\omega/2k_H)\coth(\hbar\omega/2k_B T). \tag{5.18}$$

The $(\hbar\omega/2)\coth(\hbar\omega/2k_B T)$ appearing in equations (5.17) and (5.18) has the property of approaching $k_B T$ at high temperatures ($k_B T \gg \hbar\omega/2$) so that equation (5.18) goes over into (5.16), while at low temperatures ($k_B T \ll \hbar\omega/2$) it approaches $\hbar\omega/2$ which is independent of temperature.

If one substitutes $(\hbar\omega/2)\coth(\hbar\omega/2k_B T)$ for $k_B T$ in equation (5.13) one obtains the equation derived by Hopfield (1974), and applied by him to biological electron transfer:

$$W = \frac{2\pi}{\hbar}|H_{RP}|^2(2\pi\sigma^2)^{-\frac{1}{2}} \exp\left[-\frac{(\lambda - \Delta E)^2}{2\sigma^2}\right] \tag{5.19}$$

Table 5.1. *Corresponding symbols used by different authors*

This book	Marcus (1965)	Hopfield (1974)	Jortner (1976)	G & C[†]	Ulstrup (1979)	Meaning
W, k	k, k_r, k_{unt}	W_{ab}	W	W	$W_{f\dot{v}}$	Rate constant
H_{RP}, H_{ab}	ε	T_{ab}	V_{ab}	$V \exp\{-b\sqrt{(2mV)}/\hbar\}$	V	Hamiltonian matrix element for electron transfer (eq. 2.2)
$\Delta E, -\Delta G^\circ$	$-\Delta G^{\circ\prime}$	$E_a - E_b$	ΔE	$E_l^0 - E_r^0$	$-\Delta E$	Energy loss (eq. 3.3, 5.7)
p	—	—	p	—	—	$\Delta E/\hbar\omega$ (eq. 5.20)
λ	λ	Δ	$\mu\omega^2(\Delta Q^\circ)^2/2$	—	E_r	Reorganization energy (eqs. 3.3, 5.4)
S	—	—	$S = \frac{1}{2}\Delta^2$	y_m^2	—	$\lambda/\hbar\omega$ (eq. 5.20)
ω	ω	ω	$\langle\omega\rangle$	Ω	ω	Vibrational frequency
T_c	—	$\frac{1}{2}T_a, \frac{1}{2}T_b$	—	—	—	Characteristic temperature (eq. 5.20)
k_H	—	k_a, k_b	$\mu\omega^2$	—	—	Force constant (eqs. 5.1–5.6)
k_B	—	κ	k	k	—	Boltzmann constant
$(X_B - X_A), Q$	—	$\sqrt{(x_a^2 + x_b^2)}$	ΔQ°	Δq_n	—	Displacement
$E^\ddagger$	—	—	—	ε	—	Activation energy (eqs. 3.2, 5.7)
$Q(k_H/\hbar\omega)^{\frac{1}{2}}$	—	—	Δ	—	$q_{\kappa 0}$	Dimensionless displacement
$1/k_B T, \beta$	—	—	—	—	β	Reciprocal of thermal energy

[†] Grigorov & Chenavskii (1972).

with $\sigma^2 = \hbar\omega\lambda \coth(\hbar\omega/2k_B T)$. It will also be convenient to express this equation in terms of dimensionless parameters as follows:

$$W = \frac{2\pi}{\hbar} |H_{RP}|^2 (\hbar\omega)^{-1} (2\pi\sigma_0{}^2)^{-\frac{1}{2}} \exp\left[-\frac{(S-p)^2}{2\sigma_0{}^2}\right], \tag{5.20}$$

where $\sigma_0{}^2 = S\coth(T_c/T)$; $S = \lambda/(\hbar\omega)$; $p = \Delta E/(\hbar\omega)$; and $T_c = \hbar\omega/2k_B$). A translation of the above symbols into those used by Hopfield and others is given in Table 5.1. The fit this equation makes to the data on *C. vinosum* is shown in Fig. 5.2 along with that of two other theories to be described later.

5.2.1 *Derivation by analogy to Förster–Dexter theory*

Hopefield derived his equation with an entirely different approach from that of Marcus. The approach does not consider or make use of the idea of an 'activated' or 'transition' state or of 'activation energy'. This is characteristic also of the fully quantum-mechanical theories described later. Hopfield uses the Förster (1946)–Dexter (1953) energy-transfer theory as a model for the electron-transfer process. He started with

Fig. 5.2. Comparison of theories with data on *C. vinosum* cytochrome oxidation. ----, Grigorov & Chernavskii (1972), equation (5.90), with $E^{\ddagger} = 0.152$ eV; $4S_1S_2 = 1.34 \times 10^{-6}$ and $(2\pi|H_{RP}|^2/\hbar\Gamma) = 2.24 \times 10^8$ s^{-1}. ——, Hopfield (1974), equations (5.19) or (5.29), with $\Delta E = 0.45$ eV; $T_c = 206$ K; $\lambda = 0{\cdot}090$ eV with $|H_{RP}| = 1.6 \times 10^{-3}$ eV, or $\lambda = 2.24$ eV with $|H_{RP}| = 3.6 \times 10^{-3}$ eV. – – –, Jortner (1976), equations (5.67) or (5.71), with $T_c = 278$ K; $\Delta E = 0.45$ eV; $S = 43.9$; $A = 1.14 \times 10^{12}$ s^{-1}; $|H_{RP}| = 2.4 \times 10^{-3}$ eV; and $\lambda = 2.10$ eV.

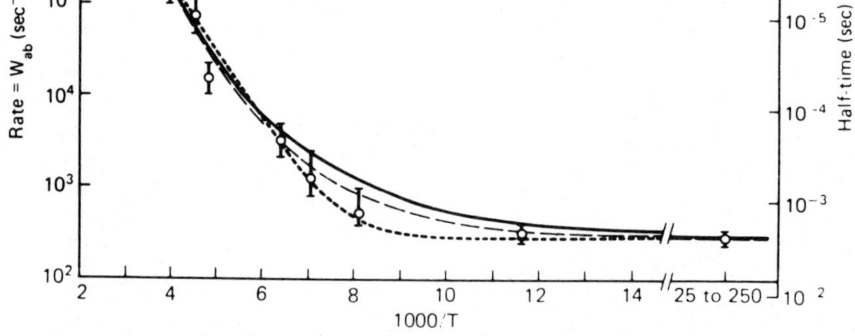

equation (2.2), which may be re-written:

$$W_{RP} = \frac{2\pi}{\hbar}|H_{RP}|^2(\mathrm{FC}), \tag{5.21}$$

where FC is an expression giving the density of states in terms of the probability of meeting the Franck–Condon conditions.

The factor FC is put together as follows: associated with the donor centre is a spectrum, $D_a(E)$, representing a distribution of energies required to remove an electron from the donor molecule together with the probabilities that the donor is in the right condition to have its electron removed with the given energy. Mathematically, the possibility that the energy of removal is between E and $E + \mathrm{d}E$ is equal to $D_a(E)\,\mathrm{d}E$. In the absence of better information, a Gaussian distribution is the most likely. It is also what one would expect if derived from a thermal distribution of coordinates of oscillators, equation (5.15) with (5.18). This distribution is represented with the probabilities horizontal and the energy axis vertical in Fig. 5.3, along with a similar distribution, D_b, for the energy given off when the acceptor receives an electron. Mathematically, the probability that the energy of insertion is between E' and $E' + \mathrm{d}E'$ is $D_b(E')\,\mathrm{d}E'$. The zero of

Fig. 5.3. Probability distributions of energy required to remove an electron from donor without change of nuclear coordinates and of the energy released when acceptor receives an electron without change of nuclear coordinates.

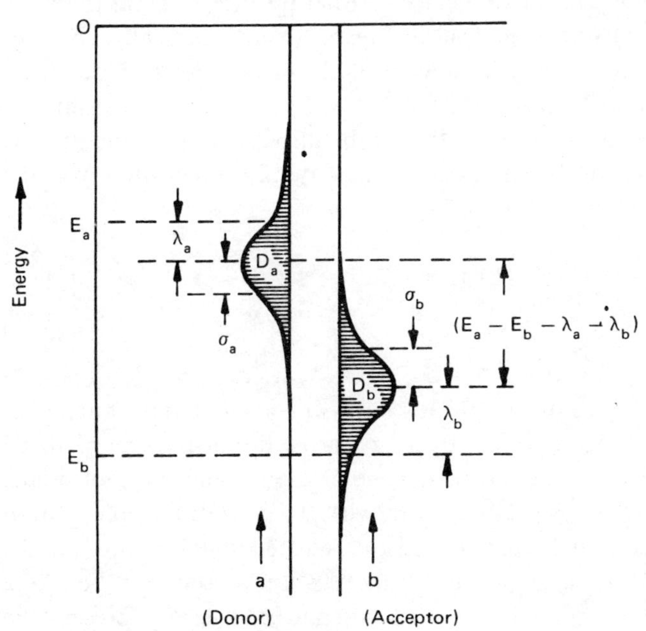

the energy scale is, in both cases, the ground state of the oxidized form. E_a is the energy of the ground state of the reduced form of the donor and E_b that of the reduced form of the acceptor. $E_a - E_b$ would be the difference in redox midpoint potentials of donor and acceptor. However, we must assume that any interaction of the charge and polarization of surroundings of the one on the other must be included so that $E_a - E_b = E_R - E_P = \Delta E$, Fig. 5.1, where E_R and E_P are energies of the entire system. D_a is centred below E_a by an amount λ_a, because ordinarily the nuclear configuration in the initial reduced state will not be an equilibrium configuration for the final, oxidized state. The transition is, therefore, to an excited state of the oxidized form and so the energy required will normally be greater than E_a. By the same reasoning D_b is displaced upward from E_b by an amount λ_b. The nuclear configuration in the oxidized form of b will correspond to an excited state of the reduced form and the energy given off will usually be less than E_b.

σ_a and σ_b designate the standard deviations of the respective Gaussian distributions. Thus:

$$D_a(E) = (2\pi\sigma_a{}^2)^{-\frac{1}{2}} \exp[-(E + E_a - \lambda_a)^2/2\sigma_a{}^2] \tag{5.22}$$

and

$$D_b(E) = (2\pi\sigma_b{}^2)^{-\frac{1}{2}} \exp[-(E + E_b + \lambda_b)^2/2\sigma_b{}^2]. \tag{5.23}$$

To conserve energy in the process, the transfer from a particular energy E in the distribution D_a must be paired with a transfer in D_b at the same energy E' that is, at the same level in Fig. 5.3. Mathematically, the probability of transition with E between E and $E + \mathrm{d}E$ and E' between E' and $E' + \mathrm{d}E'$ is $D_a(E)D_b(E')\,\mathrm{d}E\,\mathrm{d}E'$, with $E = E'$. The density of states ρ, in equation (2.2), is the amount of this probability per unit of energy, and so we divide by one of the energy intervals, say $\mathrm{d}E'$. Integration over the other gives the total density of states:

$$(\mathrm{FC}) = \int_{-\infty}^{\infty} D_a(E)D_b(E)\,\mathrm{d}E = (2\pi\sigma^2)^{-\frac{1}{2}} \exp[-(\Delta E - \lambda)^2/2\sigma^2], \tag{5.24}$$

where $\sigma^2 = \sigma_a{}^2 + \sigma_b{}^2$ and $\lambda = \lambda_a + \lambda_b$. This, with equation (5.21) is seen to be the same as equation (5.19). However, we have not yet evaluated σ.

To evaulate σ_a, assume that the energy of the donor in both oxidized and reduced state has a quadratic dependence on a nuclear coordinate, x, with the same force constant, k_a, but with the equilibrium position in the oxidized state shifted from that in the reduced state by amount Q_a. The situation is the same as in Fig. 5.1 if we let R represent not the whole system, but the donor part in the reduced state, and let P represent the oxidized state of the donor. For this, of course, P should be moved upward until the level labelled E_R there is below that labelled E_P by the amount E_a. The level E_P becomes zero and E_R becomes E_a. Equations (5.1)–(5.4) will apply here with V_R changed to V_{ra} (for reduced donor), V_P to V_{oa} (for oxidized donor), k_H to k_a, λ to λ_a, Q to Q_a, etc. Let $X - X_A = x$ and $X_B - X_A = Q_a$. Subtracting the new equation (5.1) from the new (5.2) gives the energy of removal of an electron from the reduced donor as a function of the nuclear displacement in the reduced donor:

$$E = V_{oa} - V_{ra} = \lambda_a - E_a - k_a Q_a x. \tag{5.25}$$

A similar derivation gives the energy of insertion of an electron into the oxidized acceptor as a function of the displacement, y, of the nuclei of the oxidized acceptor from its equilibrium value:

$$E = V_{ob} - V_{rb} = -\lambda_b - E_b + k_b Q_b y. \tag{5.26}$$

We assume next that the initial states of donor and acceptor are both in thermal equilibrium so that the probabilities of x and y are given by equation (5.15) with (5.18) with y substituting for x in the case of the acceptor. $D_a(E)\,dE$ is the same probability as $P(x)\,dx$ if $dE/dx = -\partial E/\partial x$ as given by equation (5.25). Thus

$$D_a(E) = -P(x)/|\partial E/\partial x|, \tag{5.27}$$

and similarly

$$D_b(E) = P(y)/|\partial E/\partial y|. \tag{5.28}$$

Solving equations (5.25) and (5.26) for x and y, respectively, and substituting for x and y in the expressions for $P(x)$ and $P(y)$, differentiating (5.25) and (5.26) to get $\partial E/\partial x$ and $\partial E/\partial y$ and substituting these into equations (5.27) and (5.28) will give equations (5.22) and (5.23) with

$$\sigma_a{}^2 = \hbar\omega_a\lambda_a \coth(\hbar\omega_a/2k_B T)$$

and

$$\sigma_b{}^2 = \hbar\omega_b\lambda_b \coth(\hbar\omega_b/2k_B T).$$

In parallel with equation (5.3), $\omega_a = (k_a/M_a)^{\frac{1}{2}}$ and $\omega_b = (k_b/M_b)^{\frac{1}{2}}$. To get equation (5.19) from (5.24) it is necessary to make the further assumptions that $\lambda_a = \lambda_b$ and $\omega_a = \omega_b = \omega$.

Hopfield's model uses two oscillators, one associated with donor and one with acceptor. The total potential energy for nuclear motion for the whole system is, therefore, $V_R = V_{ra} + V_{ob}$ in the R state and

$$V_P = V_{oa} + V_{rb}$$

in the P state. These are surfaces in a three-dimensional space and are shown as functions of both x and y in Fig. 5.4.

Fig. 5.4. Nuclear potential energy plotted against two nuclear coordinates. The nuclear coordinates x and y represent oscillator displacement in donor and in acceptor, respectively. The front face is a cross-section equivalent to Figs. 3.1 or 5.1. Three cases are shown: at the top is the 'normal' case, $\lambda > \Delta E$. In the middle is the case that $\lambda = \Delta E$. No activation is required as the R and P surfaces intersect at the equilibrium position for R. At the bottom is the 'inverted' case, $\lambda < \Delta E$. This figure produces the three cases by varying λ at constant ΔE. If one keeps λ constant and varies ΔE, one obtains the 'energy-gap law'. See equations (3.3) or (5.7). At the left are the corresponding electron insertion diagrams similar to Fig. 5.3, used by the Hopfield theory.

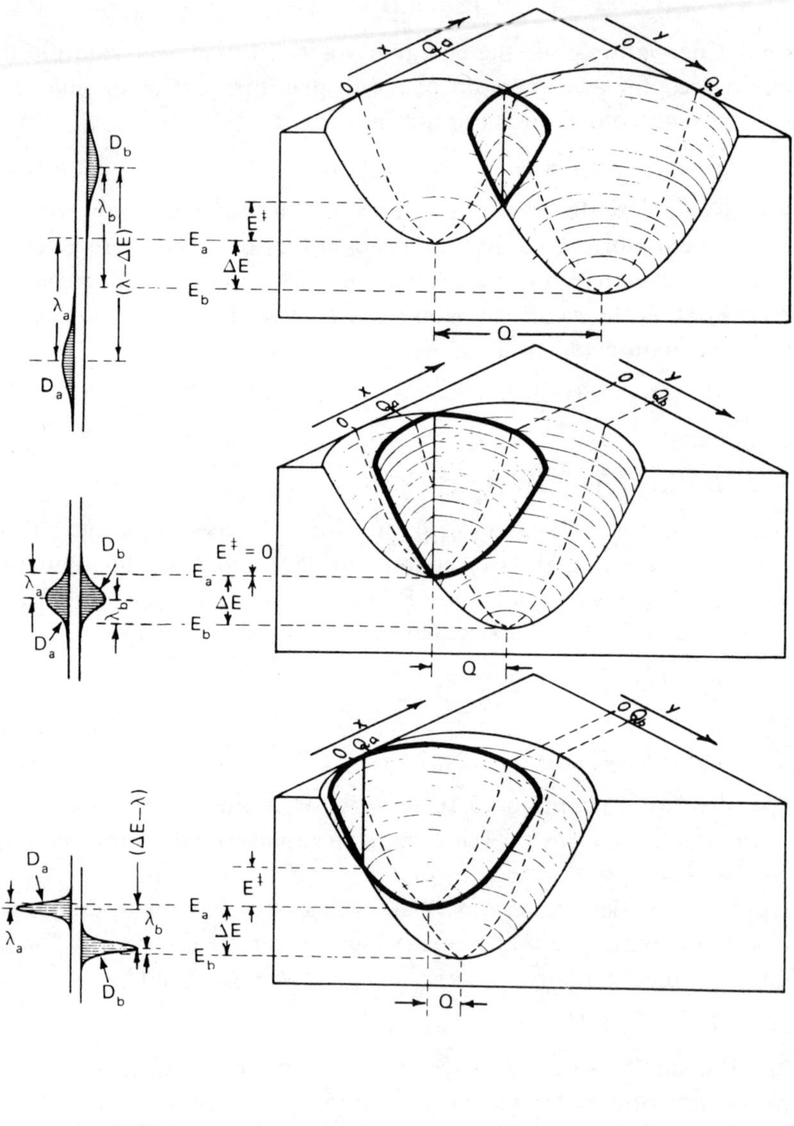

5.2.2 Calculating with equation (5.19)

Hopfield's equation is not fully quantum mechanical. However, it has the merit of approximating the temperature dependences observed in Fig. 1.1, while being relatively simple to work with. To make a fit to the rate v. temperature data one can write it in the form:

$$W_{RP} = \alpha[\tanh(T_c/T)]^{\frac{1}{2}} \exp(-\beta \tanh(T_c/T)) = W_T, \tag{5.29}$$

with

$$\alpha = (2\pi/\hbar)|H_{RP}|^2/(2\pi\hbar\omega\lambda),^{\frac{1}{2}} \tag{5.30}$$

$$\beta = (\Delta E - \lambda)^2/(2\hbar\omega\lambda), \tag{5.31}$$

$$T_c = \hbar\omega/2k_B. \tag{5.32}$$

It is seen that, for the temperature dependence, there are only three independent constants: α, β and T_c. One can evaluate these from three experimental quantities, for example, W_T and Y_T at a given high temperature T, and W_0 by the following relations:

At low temperature as T approaches 0, the hyperbolic tangent approaches 1 and:

$$W_{RP}(\text{low temp.}) \approx \alpha \exp(-\beta) = W_0. \tag{5.33}$$

The slope of an Arrhenius plot of equation (5.29) would be:

$$\frac{d \ln W_{RP}}{d(1/T)} = \frac{T_c}{\sinh(2T_c/T)} - \frac{2\beta T_c}{1 + \cosh(2T_c/T)} = Y_T. \tag{5.34}$$

The *C. vinosum* data (DeVault & Chance, 1966; DeVault *et al.*, 1967; Seibert & DeVault, 1971) may be taken as: $W_0 = 300\ \text{s}^{-1}$, $W_T = 6.9 \times 10^5\ \text{s}^{-1}$ and $Y_T = 2.53 \times 10^3$ K at $T = 294$ K. Using equations (5.29), (5.33) and (5.34) one obtains $\alpha = 1.75 \times 10^{11}\ \text{s}^{-1}$, $\beta = 20.2$ and $T_c = 206$ K. These are the parameters used in equation (5.29) to give the solid curve in Fig. 5.2.

Using the additional datum that $\Delta E = 0.45$ eV for the difference in redox midpoint potentials of the reaction centre BChl and of the low-potential cytochrome (the one observed to be oxidized in the data of Fig. 1.1) in *C. vinosum*, one can use the values of α, β and T_c to solve equations (5.32), (5.31) and (5.30) successively to obtain

$$\hbar\omega = 0.036\ \text{e}V$$

and

$$\lambda = 0.090\ \text{eV} \quad \text{and} \quad |H_{RP}| = 1.6 \times 10^{-3}\ \text{eV}$$

or

$$\lambda = 2.24\ \text{eV} \quad \text{and} \quad |H_{RP}| = 3.6 \times 10^{-3}\ \text{eV}.$$

The two values for λ and $|H_{RP}|$ result from the quadratic nature of equation (5.31). Hopfield chooses the larger value of λ on the basis that the coupling should be strong. Both are illustrated in Fig. 5.4, where it is seen that the large value (top illustration) corresponds to the 'normal' case ($\lambda > \Delta E$) while the small value (bottom illustration) corresponds to the 'inverted' case ($\Delta E > \lambda$), or the 'highly exothermic' case in the energy-gap law. These values differ a little from those given by Hopfield (1974, 1976) but his values do not fit the data well, especially at the lower temperatures and do not give the curve drawn in his figures. As already stated the above parameters give the curve drawn in Fig. 5.2.

A measure of the mismatch between $D_a(E)$ and $D_b(E)$ in units of standard deviation is the ratio $(\Delta E - \lambda)/\sigma$ which comes out 6.4 at low temperature and 4.9 at 294 K, for either value of λ. This much mismatch is required to give the observed temperature dependence.

The front faces of the (right-hand) diagrams of Fig. 5.4 are vertical cross-sections on the axes through the minima of the R and P surfaces, and correspond to the two-dimensional diagrams of Figs. 3.1 and 5.1. The equation of such an axis projected onto the x, y plane is $y = (Q_b/Q_a)x$. It goes through the saddle-point of the intersection between R and P and is the most probable reaction coordinate, although, clearly, off-axis reaction paths through higher intersection points are possible.

The line of intersection between the R and P surfaces projected on the x, y plane is given by

$$y = \frac{\lambda - \Delta E}{k_b Q_b} - \frac{k_a Q_a}{k_b Q_b} x. \tag{5.35}$$

This shows that under the assumptions used (k_a and k_b the same in both oxidized and reduced forms of their respective centres) the intersection is a straight line in the x, y plane. Under the further assumption that $k_a = k_b$ and $Q_a = Q_b$ this line is perpendicular to the front face of the diagrams in Fig. 5.4. In a proper multi-dimensional space the intersection would be a hypersurface and was treated as such in Marcus' derivations.

It is interesting that two very different viewpoints (Arrhenius activation, modified by equation (5.18), and removal and insertion of spectral overlaps) can produce the same result (equation (5.19)). From the second viewpoint the effect of temperature is on the widths, σ_a and σ_b, of the spectral functions and the resultant amount of overlap between them.

Hopfield's evaluation of the electronic matrix element, H_{RP}, as a function of electron-tunnelling distance between donor and acceptor deserves discussion, which will be left to Section 5.6.3(*a*).

Hopfield's theory is criticized by Jortner (1976), who, referring to

Soules & Duke (1971), points out that the Förster–Dexter theory should be less applicable to charge transfer than to energy transfer because of stronger coupling to vibrations in the former. In electron transfer the movement of charge over considerable distance must cause more disturbance to the surrounding molecules and even between donor and acceptor. Both donor and acceptor can be expected to be coherently coupled to many of the same vibrations in the surroundings, whereas the Förster–Dexter approach assumes that each couples to a separate set of oscillators. Further, the limiting expression obtained for the rate at low temperature differs from the fully quantum-mechanical formula. It fits the data at low temperature only by using different values for the parameters.

5.3 Quantum mechanical

The full quantum-mechanical treatment of vibration coupled to changes of electronic state has been intensively developed since about 1950. Huang & Rhys (1950) derived an equation to describe the shape of optical absorption bands of F-centres in crystals. It involved a coupling of the crystal vibrations to the electronic excitations caused by absorption of photons in the F-centres. It was restricted to a single vibrational frequency. Except for some different constants concerned with the difference between optical absorption and electron transfer, their equation is the same as that given by Levich & Dogonadze (1959, 1961) (also: Levich, 1966, 1970) for electron transfer in chemical systems and by Jortner (1976) for electron transfer in biological systems. Kubo (1952) and Lax (1952) independently developed methods of calculation, using generating functions which could handle much more general cases with a variety of oscillator frequencies and with changes of frequency considered as well as the changes of equilibrium position (Kubo & Toyozawa, 1955; Toyozawa, 1967). However, a single high frequency appeared sufficient to match the *C. vinosum* data and this reduced the more general formula to that of Huang & Rhys (Jortner, 1976). More recently Kuznetsov *et al.* (1978), Dogonadze *et al.* (1979) and Sarai (1979, 1980) have suggested the involvement of low-frequency modes, anyway, in the biological cases.

The various theories all start with the procedure originally suggested by Condon (1928) (Condon & Morse, 1929).

5.3.1 *Simple formulation*

The method is illustrated for the single-frequency case in Fig. 5.5. Fig. 5.5 is a redrawing of Figs. 3.1 and 5.1 to show the vibrational wavefunctions (squared) for the R and P systems and their overlaps. The

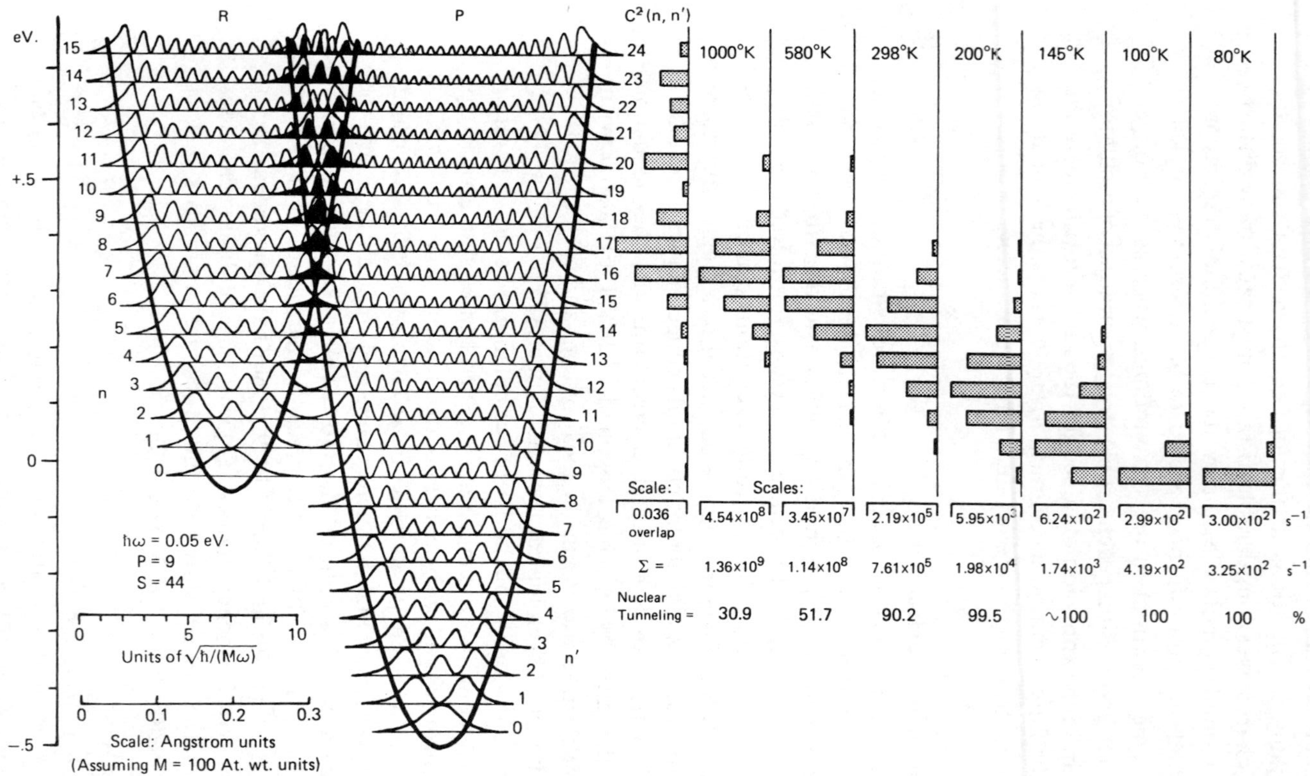

Fig. 5.5. Completely quantum-mechanical calculation of electron-transfer rates, according to equation (5.38). The diagram is similar to those of Figs. 3,1 and 5.1 drawn to a scale that matches the *C. vinsoum* data approximately. The vibrational energy levels are shown with the squares of the vibrational wave-functions superposed on them. The overlaps between those belonging to electronic state *R* and those belonging to electronic state *P* are simulated by blackening where the respective waves overlap. The correct calculation, however, of the overlap integrals, $C(n, n')$, is given by equation (5.36) and involves multiplication before squaring. The results are graphed in the column labelled $C^2(n, n')$ and are proportional to the transition rate to be expected with the corresponding values of n and n', if all R states had equal population. The other columns at the right are graphs of $C^2(n, n')$ multiplied by the population of level n in a Boltzman distribution at the temperature indicated at the top of the column. The total rates at the different temperatures are found by summing the columns and multiplying by the electronic-transition probability, A. A is chosen to be $2.26 \times 10^{12}\ s^{-1}$ to give the correct low-temperature rate for *C. vinosum* cytochrome photo-oxidation.

parameters used are close to those which fit the *C. vinosum* data. The course of reaction is as follows.

Initially the system is in state R (electron in donor) and contains any number of oscillators coupled to electron transfer all of which have frequency ω. These oscillators are treated as a single composite oscillator which can exist in quantum states whose energies differ from each other by multiples of $\hbar\omega$ and are plotted in Fig. 5.5 with designations at the left of the figure running from '0' for the lowest to '15' for the highest plotted. These designations are values of the quantum number n. The vibrational energy in state n is $E_n = (n + \frac{1}{2})\hbar\omega$. At a given temperature T, the probability that an oscillator is in state n is proportional to the Boltzmann factor $\exp(-E_n/k_B T)$. In Fig. 5.5 on top of each energy level is drawn a wavy curve, the square of the wave-function for that state, proportional to the probability of finding the coordinate of oscillation, x, at the value of x given by the abscissa if the oscillator is known to be in the given state.

At some infinitesimal moment the electron suddenly tunnels from donor to acceptor and the system finds itself in the P state. The change takes place without loss or gain of energy and so the system will have the same level in Fig. 5.5 after the transition as before but will be shifted to the right, to the P curve. If the quantum number, n, of the oscillator was 10 in state R it will be $n' = 19$ in state P. In general, $n' = n + p$, where $p = \Delta E/\hbar\omega$. The additional vibrational energy is just equal to the loss of electronic energy resulting from the electron transfer. Later the extra vibrational energy will dissipate as heat while the P system settles to a new thermal equilibrium.

The probability of the electron transfer and, therefore, the rate of reaction depends not only on the ease of the electron tunnelling itself but also on the willingness of the oscillators to make the necessary accompanying quantum changes. The latter is proportional to the square of the 'overlap integral', $C(n, n')$, between the vibrational state n' in state P and the corresponding state n in R. This integral is defined as:

$$C(n, n') = \int \chi_n \chi_{n'} \, dx, \tag{5.36}$$

where χ_n is the vibrational wave-function for state n in R, $\chi_{n'}$ is the wave-function for state n' in P and x is the oscillator coordinate. Its squared absolute value, $C^2(n, n')$, is called the Franck–Condon factor and is the true representation of the Franck–Condon principle. The statement that the principle requires a vertical transition from the R curve to the P curve is seen to be only approximate. In Fig. 5.5, $C^2(n, n')$ is represented qualitatively by the shaded areas of overlap of the probability functions for R

and P at each energy level, but, of course, the actual calculation involves multiplication of their square roots as stated by equation (5.36). This means that neighbouring areas of overlap will have opposite sign and will tend to cancel each other. The following exact formula for $C^2(n, n')$ with $n' = n + p$ for the case that the frequency does not change can be derived from formulas for overlap integrals given by Manneback (1951):

$$C^2(n, n+p) = n!(n+p)!S^p e^{-S}\left[\sum_{i=0}^{n} \frac{(-S)^i}{i!(n-i)!(p+i)!}\right]^2. \tag{5.37}$$

As already stated, $S = \lambda/(\hbar\omega)$ and $p = \Delta E/(\hbar\omega)$. Relative values of $C^2(n, n')$ are plotted as horizontal bars at the level corresponding to the respective n and n' in the column labelled '$C^2(n, n')$' just to the right of the R, P diagram in Fig. 5.5. These are calculated with $S = 44$ and $p = 9$. The longest bar ($n = 8$) has the value 0.036. The bar for $n = 10$ is small because of interference (cancellation) between the three main overlap areas. For a given value of electron-tunnelling probability, $C^2(n, n')$ is proportional to the rate of reaction exhibited by a system known to be initially in state n.

To get the overall rate of reaction it is only necessary to multiply each $C^2(n, n')$ by the fraction of systems expected to be in state n at a given temperature, to sum the products over all values of n and multiply by the probability of the electron tunnelling. Thus we have:

$$W = A \sum_{n=0}^{\infty} C^2(n, n+p)B(n, T), \tag{5.38}$$

where the constant, A $(= 2\pi|H_{ab}|^2/\hbar^2\omega)$, represents the part of the rate dependent on the electron-tunnelling probability and B gives the fraction of oscillators in state n at temperature T:

$$B(n, T) = e^{-\beta\hbar\omega n}(1 - e^{-\beta\hbar\omega}) \tag{5.39}$$

with $\beta = 1/k_B T$. Equation (5.38) is seen to be the same as equation (5.21) with:

$$(\text{FC}) = \sum C^2(n, n+p)B(n, T)/\hbar\omega. \tag{5.40}$$

The quantity $\hbar\omega$ is the separation between states to which the electron can tunnel and its reciprocal is thus the density of states, ρ, equation (2.2), after correction by the Franck–Condon factors, which are in turn corrected by the Boltzmann probabilities.

The columns beyond that for $C^2(n, n')$ in Fig. 5.5 exhibit $C^2(n, n+p)$, given in the first column, multiplied each by the corresponding $B(n, T)$ for the value of T given at the head of each column. 0.05 eV was used for the value of $\hbar\omega$. The scale is different for each column, being adjusted to give the same length of bar for the longest bar in each. The scale is given below each column in reciprocal seconds for $A = 2.26 \times 10^{12}\ \mathrm{s}^{-1}$. The row below the scales gives the sum of the terms in the column above each value and thus the total predicted rate for the given temperature.

The cross-over between the R and P curves occurs between $n = 6$ and $n = 7$. If we consider that all values of n below 7 correspond to nuclear tunnelling, then the rate of nuclear tunnelling is seen to be the sum of the terms for $n = 0$–6. The percentage of the total rate which is included in this sum is given by the bottom row of figures. It is seen that at room temperature 90% of the rate involves nuclear tunnelling and the percentage increases as the temperature drops. At the lowest temperatures the situation is approached in which only $n = 0$ is populated and all of the reaction is through $n = 0$. The rate then becomes completely temperature-independent. Between 80 and 200 K the reaction is almost entirely by temperature-dependent nuclear tunnelling. The barrier to nuclear tunnelling involved in the case $n = 6$ is so miniscule that one is tempted not to include $n = 6$ in nuclear tunnelling. This would drop the room temperature nuclear tunnelling to 70%.

The columns for 580 and 1000 K (not attainable in actual biological systems, but of theoretical interest) show that at the higher temperatures the most active values of n are those close to the value at the crossing point of the R and P curves. This is the result of the competition between larger overlap integrals above the crossing point v. the larger populations at the lower values of n. It is the quantum-mechanical approximation to Arrhenius behaviour which postulates that all reaction occurs by excitation to the crossing point. We thus see that the activation energy, 0.18 eV (Table 4.2), derived from the slope of the temperature-dependent portion of the Arrhenius plot, Fig. 1.1, actually corresponds more to temperature-dependent nuclear tunnelling and the true Arrhenius activation energy is more like 0.33 eV and would only be found at much higher temperatures.

5.3.2 *General formulations*

The foregoing procedure becomes too complicated to handle in the case of many modes with different frequencies of vibration. The formulations to be described next do precisely the same job but in a more tractible manner. I will attempt to sketch a derivation of the Kubo (1952) formulation to the minimum extent necessary to understand the presentation by Jortner (1976). My understanding, such as it is, was aided considerably by the expositions of Sarai (1979, 1980), Ulstrup (1979) and Kubo & Toyozawa (1955). I will take the single-frequency case as an example leading to the Huang–Rhys formula.

Equations (5.21) and (5.38) already assume the 'Condon approximation'

(Condon, 1928), which is that the electronic matrix element is sufficiently constant in the cross-over region, nuclear configuration C or X_C in Figs. 3.1, 3.2 and 5.1, that it can be factored out of the total matrix element leaving the nuclear part in the factor (FC). The idea is that the nuclear vibrations with amplitudes of the order of 0.1 Å can have little effect on the purely electronic factor influencing the probability of the electron tunnelling across several Å. The assumption that the electronic parts can be considered separately from the nuclear parts is the 'Born–Oppenheimer approximation' (Born & Oppenheimer, 1927; Born & Huang, 1954). Further consideration of this aspect is left to Section A.2, in the Appendix. See also Section 5.5.2.

Equation (5.36) can be used to rewrite equation (5.40) for the factor (FC) as follows:

$$(\mathrm{FC}) = \sum_{n=0}^{\infty} \sum_{n'=0}^{\infty} \int \chi_n(x)\chi_{n'}(x')\,\mathrm{d}x \int \chi_n(y)\chi_{n'}(y')\,\mathrm{d}y B(n, T) \times \delta(E_{n'} - E_n - \Delta E). \tag{5.41}$$

Here x' is the same as x except that the origin (= equilibrium position) is displaced by amount Q. The symbols y and y' are the same as x and x' respectively, but are different in order to distinguish to which integrals the factors belong after they are mixed up by further manipulations. $E_{n'}$ is the vibrational energy in the final state and is equal to $n'\hbar\omega$ (putting the zero point energy into that attributed to the electronic state). E_n, equal to $n\hbar\omega$, is the initial vibrational energy. ΔE is the loss in electronic energy. The (Dirac) δ-function could just as well be written: $\delta(n' - n - p)/\hbar\omega$. It and the summation over n' are not necessary in the single-frequency case because one then simply sets $n' = n + p$ to conserve energy and sums over n only. However, in the multiple-frequency cases handled by the expanded theories there is a factor $\{\Sigma_n \Sigma_{n'} C^2(n, n')B(n, T)\}$ for each mode of oscillation. The result is equivalent to the sum over all possible combinations of the n and n' for each factor, of the products of the factors in sets, each set containing a factor for each mode. This is then multiplied by a single δ-function in which $E_{n'}$ is replaced by a sum of the $E_{n'}$ for each factor in a set and likewise E_n by the sum of the E_n. The δ-function then picks out only those sets having combinations of the n and n' which conserve energy. As mentioned by Jortner (1976), the δ-function should actually be a Lorentzian function to allow for a width resulting from finite lifetimes.

The δ-function is transformed to an integral by the following:

$$\delta(n' - n - p) = \frac{\omega}{2\pi} \int_{-\infty}^{\infty} \mathrm{e}^{i(n' - n - p)\omega t}\,\mathrm{d}t. \tag{5.42}$$

Using equations (5.42) and (5.39) we can rearrange equation (5.41) to

$$(\mathrm{FC}) = \frac{1}{2\pi\hbar}(1 - \mathrm{e}^{-\beta\hbar\omega}) \int \mathrm{d}t \int \mathrm{d}x \int \mathrm{d}y \sum_{n=0}^{\infty} \chi_n(x)\chi_n(y)\mathrm{e}^{n(\beta\hbar\omega + i\omega t)}$$
$$\times \sum_{n'=0}^{\infty} \chi_{n'}(x')\chi_{n'}(y')\mathrm{e}^{n' i\omega t}\cdot\mathrm{e}^{-pi\omega t}. \tag{5.43}$$

The two summations can each be handled by application of Mehler's formula (Bateman (1953), p. 194, equation 22):

$$\sum_{r=0}^{\infty} \frac{(Z/2)^r}{r!} H_r(\xi)H_r(\eta) = \frac{1}{(1 - Z^2)^{\frac{1}{2}}} \exp\left[\frac{2\xi\eta Z - (\xi^2 + \eta^2)Z^2}{1 - Z^2}\right], \tag{5.44}$$

where H_r is the Hermite polynomial of order r. It becomes the harmonic oscillator wave-function, χ_r, by multiplication by $\mathrm{e}^{-\xi^2/2}$ or $\mathrm{e}^{-\eta^2/2}$, depending upon which argument is used, and by the normalizing factor, N_r. For normalization by integration over $\mathrm{d}\xi$ or $\mathrm{d}\eta$:

$$N_r = (\pi^{\frac{1}{2}} 2^r r!)^{-1/2}. \tag{5.45}$$

We see that all but the $\pi^{\frac{1}{2}}$ is already built into the left side of equation (5.44). Multiplying both sides of (5.44) by $\pi^{-1/2}\exp[-\frac{1}{2}(\xi^2 + \eta^2)]$ gives:

$$\sum_{r=0}^{\infty} Z^r \chi_r(\xi)\chi_r(\eta) = \frac{1}{[\pi(1 - Z^2)]^{\frac{1}{2}}} \exp\left\{\frac{2\xi\eta Z - \frac{1}{2}(\xi^2 + \eta^2)(1 + Z^2)}{1 - Z^2}\right\}, \tag{5.46}$$

This is sometimes called a 'Slater sum' (Rice, 1967).

In applying equation (5.46) to the summation over n in equation (5.43), let $r = n$, $\xi = x(M\omega/\hbar)^{\frac{1}{2}}$, $\eta = y(M\omega/\hbar)^{\frac{1}{2}}$, and $Z = \exp[-(\beta\hbar + it)\omega]$. The χ in equation (5.43) are normalized against $\mathrm{d}x$ or $\mathrm{d}y$ and their normalization constants, N_n, are $(M\omega/\hbar)^{\frac{1}{4}}$ times that in equation (5.45). Therefore, both sides of equation (5.44) must be multiplied by $(M\omega/\hbar)^{\frac{1}{2}}$ to substitute for the summation over n in equation (5.43). The summation over n then becomes:

$$\frac{(M\omega/\pi\hbar)^{\frac{1}{2}}}{\{1 - \exp[-2(\beta\hbar + it)\omega]\}^{\frac{1}{2}}}$$
$$\times \exp\left\{\frac{2xy\,\mathrm{e}^{-(\beta\hbar + it)\omega} - \frac{1}{2}(x^2 + y^2)(1 + \mathrm{e}^{-2(\beta\hbar + it)\omega})}{1 - \exp[-2(\beta\hbar + it)\omega]}\left(\frac{M\omega}{\hbar}\right)\right\}. \tag{5.47}$$

Next we change variables: let $v = \frac{1}{2}(x + y)$ and $w = x - y$. Then xy becomes $v^2 - \frac{1}{4}w^2$ and $x^2 + y^2$ becomes $2(v^2 + \frac{1}{4}w^2)$. With some algebraic manipulation expression (5.47) becomes:

$$\frac{(M\omega/\pi\hbar)^{\frac{1}{2}}}{\{1-\exp[-2(\beta\hbar+it)\omega]\}^{\frac{1}{2}}}$$
$$\times \exp\left\{\left[-v^2\tanh\left(\frac{(\beta\hbar+it)\omega}{2}\right)\right.\right.$$
$$\left.\left.-\tfrac{1}{4}w^2\coth\left(\frac{(\beta\hbar+it)\omega}{2}\right)\right]\frac{M\omega}{\hbar}\right\}. \tag{5.48}$$

To apply equation (5.46) to the sum over n' in equation (5.43) let $r = n'$, $\xi = (x-Q)(M\omega/\hbar)^{\frac{1}{2}}$, $\eta = (y-Q)(M\omega/\hbar)^{\frac{1}{2}}$ and $Z = e^{i\omega t}$. The factor $(M\omega/\hbar)^{\frac{1}{2}}$ for the change in normalization constants will apply again and the summation becomes:

$$\frac{(M\omega/\pi\hbar)^{\frac{1}{2}}}{[1-\exp(2i\omega t)]^{\frac{1}{2}}}$$
$$\times \exp\left\{\frac{2(x-Q)(y-Q)\,e^{i\omega t}-\frac{1}{2}[(x-Q)^2+(y-Q)^2](1+e^{2i\omega t})}{1-\exp(2i\omega t)}\left(\frac{M\omega}{\hbar}\right)\right\}. \tag{5.49}$$

We make the same change of variable as above: $v = \frac{1}{2}(x+y)$ and $w = x-y$ so that $(x-Q)(y-Q)$ becomes $[(v-Q)^2 - \frac{1}{4}w^2]$ and $(x-Q)^2 + (y-Q)^2$ becomes $2[(v-Q)^2 + \frac{1}{4}w^2]$. Expression (5.49) is thereby changed, with further manipulation, to:

$$\frac{(M\omega/\pi\hbar)^{\frac{1}{2}}}{[1-\exp(2i\omega t)]^{\frac{1}{2}}}$$
$$\times \exp\left\{\left[-(v-Q)^2\tanh\left(-\frac{i\omega t}{2}\right)-\tfrac{1}{4}w^2\coth\left(-\frac{i\omega t}{2}\right)\right]\frac{M\omega}{\hbar}\right\}. \tag{5.50}$$

The integrals over dx and dy in equation (5.43) now become integrals over dv and dw. The scales are such that $dx\,dy = dv\,dw$. Equation (5.43) can now be written:

$$(\mathrm{FC}) = \frac{(1-e^{-\beta\omega\hbar})}{2\pi\hbar}\int dt\frac{(M\omega/\pi\hbar)\exp(-pi\omega t)}{\{1-\exp[-2(\beta\hbar+it)\omega]\}^{\frac{1}{2}}[1-\exp(2i\omega t)]^{\frac{1}{2}}}$$
$$\times \int dv\exp\left\{\left[-v^2\tanh\left(\frac{(\beta\hbar+it)\omega}{2}\right)-(v-Q)^2\tanh\left(-\frac{i\omega t}{2}\right)\right]\frac{M\omega}{\hbar}\right\}$$
$$\times \int dw\exp\left\{\left[-w^2\coth\left(\frac{(\beta\hbar+it)\omega}{2}\right)-w^2\coth\left(-\frac{i\omega t}{2}\right)\right]\frac{M\omega}{4\hbar}\right\}. \tag{5.51}$$

The integrals over dv and dw are now clearly seen as Gaussians and can be handled in the usual way. That in w is seen immediately to equal:

$$\pi^{\frac{1}{2}}\left\{\frac{M\omega}{4\hbar}\left[\coth\left(\frac{(\beta\hbar+it)\omega}{2}\right)+\coth\left(-\frac{i\omega t}{2}\right)\right]\right\}^{-\frac{1}{2}} \tag{5.52}$$

The integral over dv requires more manipulation but comes out:

$$\pi^{\frac{1}{2}}\left\{\frac{M\omega}{\hbar}\left[\tanh\left(\frac{(\beta\hbar+it)\omega}{2}\right)+\tanh\left(\frac{-i\omega t}{2}\right)\right]\right\}^{-\frac{1}{2}}$$
$$\times\exp\left\{-Q^2\left[\frac{\tanh[\frac{1}{2}(\beta\hbar+it)\omega]\tanh[\frac{1}{2}(-i\omega t)]}{\tanh[\frac{1}{2}(\beta\hbar+it)\omega]+\tanh[\frac{1}{2}(-i\omega t)]}\right]\frac{M\omega}{\hbar}\right\}. \tag{5.53}$$

Expressions (5.52) and (5.53) can be substituted into (5.51), but first notice that the factors:

$$\left\{\coth\left(\frac{(\beta\hbar+it)\omega}{2}\right)+\coth\left(\frac{-i\omega t}{2}\right)\right\}^{-\frac{1}{2}}$$
$$\times\left\{\tanh\left(\frac{(\beta\hbar+it)\omega}{2}\right)+\tanh\left(\frac{-i\omega t}{2}\right)\right\}^{-\frac{1}{2}}$$
$$=\left\{\frac{\sinh(\frac{1}{2}\beta\hbar\omega)}{\sinh[\frac{1}{2}(\beta\hbar+it)\omega]\sinh\frac{1}{2}(-i\omega t)}\right\}^{-\frac{1}{2}}$$
$$\times\left\{\frac{\sinh(\frac{1}{2}\beta\hbar\omega)}{\cosh[\frac{1}{2}(\beta\hbar+it)\omega]\cosh[\frac{1}{2}(-i\omega t)]}\right\}^{-\frac{1}{2}}$$
$$=\left\{\sinh\left(\frac{\beta\hbar\omega}{2}\right)\right\}^{-1}\{\tfrac{1}{4}\sinh[(\beta\hbar+it)\omega]\sinh(-i\omega t)\}^{\frac{1}{2}}$$
$$=(\tfrac{1}{2}e^{\frac{1}{2}\beta\hbar\omega})^{-1}(1-e^{-\beta\hbar\omega})^{-1}\{\tfrac{1}{16}e^{(\beta\hbar+it)\omega}[1-e^{-2(\beta\hbar+it)\omega}]e^{-i\omega t}$$
$$\times[1-e^{2i\omega t}]\}^{\frac{1}{2}}$$
$$=\tfrac{1}{2}(1-e^{-\beta\hbar\omega})^{-1}[1-e^{-2(\beta\hbar+it)\omega}]^{\frac{1}{2}}[1-e^{2i\omega t}]^{\frac{1}{2}}. \tag{5.54}$$

These factors cancel corresponding factors in equation (5.51) which becomes:

$$(\mathrm{FC})=\frac{1}{2\pi\hbar}\int \mathrm{d}t$$
$$\times\exp\left\{-Q^2\left[\frac{\tanh[\frac{1}{2}(\beta\hbar+it)\omega]\tanh[\frac{1}{2}(-i\omega t)]}{\tanh[(\beta\hbar+it)\omega]+\tanh[\frac{1}{2}(-i\omega t)]}\right]\frac{M\omega}{\hbar}-pi\omega t\right\}$$

$$= \frac{1}{2\pi\hbar} \int \mathrm{d}t \exp\left\{ -Q^2 \frac{M\omega}{\hbar} \left[\frac{\sinh(\frac{1}{2}(\beta\hbar + it)\omega)\sinh(-i\omega t/2)}{\sinh(\beta\hbar\omega/2)} \right] - pi\omega t \right\}. \tag{5.55}$$

Manipulation of the hyperbolic sines yields:

$$(\mathrm{FC}) = \frac{1}{2\pi\hbar} \int \mathrm{d}t$$

$$\times \exp\left\{ -Q^2 \frac{M\omega}{2\hbar} \left[\frac{1 - \mathrm{e}^{-(\beta\hbar + it)\omega} - \mathrm{e}^{i\omega t} + \mathrm{e}^{-\beta\hbar\omega}}{1 - \mathrm{e}^{-\beta\hbar\omega}} \right] - pi\omega t \right\}. \tag{5.56}$$

We have already defined $Q^2 M\omega/2\hbar$ as S. See Equations (5.3), (5.4) and under Equation (5.20). If the system consists of a group of oscillators all having the same frequency, S is a summation of values for each oscillator. These can differ from each other because of different degrees of coupling to the electron transfer. The average quantum number, $\bar{n}$, in an ensemble of identical oscillators at thermal equilibrium is $(\mathrm{e}^{\beta\hbar\omega} - 1)^{-1}$. Thus $\mathrm{e}^{\beta\hbar\omega} = (\bar{n} + 1)/\bar{n}$. Substituting these into (5.56) gives:

$$(\mathrm{FC}) = \frac{1}{2\pi\hbar} \int_{-\infty}^{\infty} \mathrm{d}t \exp\{ -S[2\bar{n} + 1 - (\bar{n} + 1)\,\mathrm{e}^{i\omega t} - \bar{n}\,\mathrm{e}^{-i\omega t}] - pi\omega t \}. \tag{5.57}$$

This is the form, suitable for harmonic oscillators which do not change frequency, of the generating function, developed first by Kubo (1952) and Lax (1952), which is the starting point for Jortner's (1976) presentation. It is usually written:

$$W = \frac{2\pi}{\hbar} |H_{RP}|^2 \frac{1}{2\pi\hbar} \int_{-\infty}^{\infty} f(t)\,\mathrm{e}^{-i\Delta E t\,\hbar}\,\mathrm{d}t, \tag{5.58}$$

where

$$f(t) = \exp\{ -G + G_+(t) + G_-(t) \},$$
$$G_+(t) = \sum_k S_k(\bar{n}_k + 1)\,\mathrm{e}^{i\omega_k t},$$
$$G_-(t) = \sum_k S_k \bar{n}_k\,\mathrm{e}^{-i\omega_k t},$$
$$G = G_+(0) + G_-(0).$$

The subscripts k designate the different modes in a multi-oscillator system and the summations are over all modes. The t need not represent 'time' as it merely plays the role of a dummy in the integral representation of the δ-function. Jortner's 't' is actually our $t/\hbar$. Ulstrup (1979) uses 'α' or 'θ' (Dogonadze *et al.*, 1979) corresponding to our $-it/\beta\hbar$.

G_+ represents absorption of one vibrational quantum by each oscillator and G_- release of one. The exponential of the Gs forms a sum of all possible combinations of loss and gain of vibrational energy.

The factor $\exp[G_+(t) + G_-(t)]$ can be written:

$$\exp\left\{\sum_k S_k[\bar{n}_k(\bar{n}_k+1)]^{\frac{1}{2}}\left[\left(\frac{\bar{n}_k+1}{\bar{n}_k}\right)^{\frac{1}{2}} e^{i\omega_k t} + \left(\frac{\bar{n}_k}{\bar{n}_k+1}\right)^{\frac{1}{2}} e^{-i\omega_k t}\right]\right\}. \tag{5.59}$$

This has a form to which the formula† (Watson, 1945, p.14, with $z \to ix$, $t \to iy$):

$$\exp\left\{\frac{x}{2}\left(y + \frac{1}{y}\right)\right\} = \sum_{n=-\infty}^{\infty} y^n I_n(x) \tag{5.60}$$

can be applied directly, giving

$$\prod_k \left\{\sum_{m_k=-\infty}^{\infty}\left(\frac{\bar{n}_k+1}{\bar{n}_k}\right)^{m_k/2} I_{m_k}(2S_k[\bar{n}_k(\bar{n}_k+1)]^{\frac{1}{2}})\, e^{m_k i\omega_k t}\right\}. \tag{5.61}$$

$I_n(x)$ is the modified Bessel function, defined as:

$$I_p(Z) = \sum_{k=0}^{\infty} \frac{(Z/2)^{p+2k}}{k!(p+k)!}. \tag{5.62}$$

The net change in quantum number for oscillator mode k is m_k. Expression (5.61) is in the form of a product of sums but can also be seen as a sum of products each product having a different combination of values for the m_k. Within each product one may collect the exponential factors and multiply them by $\exp(-i\Delta Et/\hbar)$, common to all the products (equation (5.58)) to give:

$$\exp\{\sum_k i m_k \omega_k t - i\Delta Et/\hbar\}. \tag{5.63}$$

When this is integrated against dt, one gets $2\pi\delta(\Sigma m_k\omega_k - \Delta E/\hbar)$, which is zero unless the product is of factors with a combination of m_k such that energy is conserved.

A variation of procedure is to exclude the mode, $k = s$, with lowest frequency, ω_s, or a group of such modes all with the same frequency, ω_s, from the product over k in expression (5.61). If s is a group, it is treated as a single mode with composite S_s. One collects the exponentials from the factors in the product over k and multiplies by

$$\exp(-i\Delta Et/\hbar)$$

† Pointed out to me by A Sarai.

to give, in place of (5.63)

$$\exp\Big(\sum_{k\neq s} im_k\omega_k t - i\Delta Et/\hbar + ip\omega_s t\Big), \tag{5.64}$$

where m_s is replaced by p. The integral of (5.64) over dt can be written

$$\frac{2\pi}{\omega_s}\delta\left(\sum_{k\neq s} m_k\frac{\omega_k}{\omega_s} - \frac{\Delta E}{\hbar\omega_s} + p\right),$$

which ensures that

$$p(m) = (\Delta E/\hbar - \sum_{k\neq s} m_k\omega_k)/\omega_s. \tag{5.65}$$

With this as a definition for p, the δ-function can be dropped. The factor $2\pi/\omega_s$ is retained and, with the $\frac{1}{2}\pi\hbar$ in Equation (5.58) gives the density of states appropriate to the steps of energy, $\hbar\omega_s$, available for the final balancing of energy via adjustment of p. The value of p (or m_s) is thus fixed by the values of the other m_k to ensure energy conservation and summation over m_s is obviated.

5.3.3 *Special cases*

For example, suppose there are two modes: ω_s and one of higher frequency, ω_k. Since k has only one value the products and sum over k become unnecessary. Then Equation (5.58) with (5.61) and (5.64) becomes

$$\begin{aligned} W = {} & \frac{2\pi}{\hbar}|H_{RP}|^2\frac{1}{2\pi\hbar}\exp(-G)\sum_m\left(\frac{\bar{n}_k+1}{\bar{n}_k}\right)^{\frac{1}{2}m} I_m(2S_k[\bar{n}_k(\bar{n}_k+1)]^{\frac{1}{2}}) \\ & \times\left(\frac{\bar{n}_s+1}{\bar{n}_s}\right)^{\frac{1}{2}p} I_p(2S_s[\bar{n}_s(\bar{n}_s+1)]^{\frac{1}{2}}) \\ & \times\int \mathrm{d}t\,\exp\{i(m\omega_k + p\omega_s - \Delta E/\hbar)t\} \\ = {} & \frac{2\pi}{\hbar^2\omega_s}|H_{RP}|^2\exp(-G)\sum_m\left(\frac{\bar{n}_k+1}{\bar{n}_k}\right)^{\frac{1}{2}m} I_m(2S_k[\bar{n}_k(\bar{n}_k+1)]^{\frac{1}{2}}) \\ & \times\left(\frac{\bar{n}_s+1}{\bar{n}_s}\right)^{\frac{1}{2}p} I_p(2S_s[\bar{n}_s(\bar{n}_s+1)]^{\frac{1}{2}}), \end{aligned} \tag{5.66}$$

with $p(m) = (\Delta E/\hbar - m\omega_k)/\omega_s$ and $G = S_k(2\bar{n}_k+1) + S_s(2\bar{n}_s+1)$. This is Jortner's (1976) equation (3*a*).

Returning to the single-mode case, ω_s becomes the only frequency and the corresponding factor in (5.61) the only factor, and so, dropping the subscript, $p = \Delta E/\hbar\omega$, as defined earlier.

Equation (5.58) or equation (5.21) with (5.57) then becomes

$$W = \frac{2\pi}{\hbar^2 \omega} |H_{RP}|^2 \, e^{-S(2n+1)} \left(\frac{\bar{n}+1}{\bar{n}} \right)^{\frac{1}{2}p} I_p(2S[\bar{n}(\bar{n}+1)]^{\frac{1}{2}}). \qquad (5.67)$$

This is Jortner's (1976) equation (4), corrected, applied by him to the *C. vinosum* data. It is plotted in Fig. 5.12 as calculated with the parameters that give the best fit.

Equation (5.67) gives numerical results identical to those given by equations (5.37)–(5.39). The latter are about as easy to calculate in the single-frequency case but the generating function used by the former is designed to handle much more easily the multi-frequency case.

At very low temperature, $\bar{n} \to 0$, and equation (5.67) becomes

$$W_0 = \frac{2\pi}{\hbar^2 \omega} |H_{RP}|^2 \, e^{-S} S^p \frac{1}{p!}. \qquad (5.68)$$

At very high temperature, Z becomes large and $I_p(Z)$ can be approximated with an asymptotic expression (Watson, 1945, p. 203):

$$I_p(Z) \approx (2\pi Z)^{-\frac{1}{2}} \exp(Z - p^2/2Z). \qquad (5.69)$$

Substituting this, $(\bar{n}+1)/\bar{n} = \exp(\hbar\omega/k_B T)$, and the following high-temperature, limiting values:

$$2S[\bar{n}(\bar{n}+1)]^{\frac{1}{2}} - S(2\bar{n}+1) \to -S\hbar\omega/4k_B T$$

and

$$2S[\bar{n}(\bar{n}+1)]^{\frac{1}{2}} \to 2Sk_B T/\hbar\omega$$

into equation (5.67) gives

$$W \approx \frac{2\pi}{\hbar} |H_{RP}|^2 (4\pi\lambda k_B T)^{-\frac{1}{2}} \exp\left\{ -\frac{(\Delta E - \lambda)^2}{4\lambda k_B T} \right\}. \qquad (5.70)$$

This is seen to be the same as equation (5.13) and so the fully quantum-mechanical approach, as well as the semiclassical, reproduces the classical Arrenius activation behaviour at high temperature without having postulated a necessity to go over a barrier.

5.3.4 *Calculating with equation* (5.67)

In order to compare with data on temperature dependence it is convenient to put equation (5.67) into the form:

$$W_T = AI_p(Z) \exp[(pT_c/T) - S\coth(T_c/T)], \qquad (5.71)$$

where $T_c = \hbar\omega/2k_B$, $Z = S/\sinh(T_c/T)$ and $A = 2\pi |H_{RP}|^2/\hbar^2\omega$. Differentiating with respect to T gives for $Y_T = -\,\mathrm{d}(\ln W_T)/\mathrm{d}(1/T)$:

$$Y_T = T_C\left\{\frac{S\coth(T_c/T)}{\sinh(T_c/T)}\frac{I_{p+1}(Z)}{I_p(Z)} + p\left[\coth\left(\frac{T_c}{T}\right) - 1\right] - \frac{S}{\sinh^2(T_c/T)}\right\}. \quad (5.72)$$

Dividing (5.71) by (5.68) gives

$$\frac{W_T}{W_0} = p!\,I_p(Z)\exp\left\{\frac{pT_c}{T} - S\coth\left(\frac{T_c}{T}\right) + S - p\ln S\right\}. \quad (5.73)$$

Equations (5.72) and (5.73) may be solved simultaneously for S and p if one has data giving Y_T, the (negative) slope of an Arrhenius plot at temperature T, W_T, the rate at the same temperature, W_0, the constant rate at very low temperature, and ΔE. T_c is replaced by $\Delta E/2pk_B$. To solve these equations one can, with each, find S as a function of p by trying different values of S for a given value of p until consistency is obtained. The Bessel functions are easily programmed, even on a hand computer. Then for each equation plot S v. p. Where the two curves cross is the solution for S and p. Although p is required by the theory to be an integer, the model used is too simple to fit experimental data accurately and it is more convenient to use continuous values. For this the definition of the modified Bessel function is

$$I_p(Z) = \sum_{k=0}^{\infty}\frac{(Z/2)^{p+2k}}{k!\,\Gamma(p+k+1)} \quad (5.74)$$

and $p!$ in equations (5.68) and (5.73) becomes $\Gamma(p+1)$. Using the same experimental data for *C. vinosum* as given in connexion with the Hopfield formula one obtains $S = 43.9$ and $p = 9.39$. Substituting these into equation (5.68) gives $A = 1.16 \times 10^{12}\ \mathrm{s}^{-1}$ and $|H_{RP}| = 2.4 \times 10^{-3}$ eV. ΔE was taken as 0.45 eV. Other parameters calculated from these are $T_c = 278$ K, $\hbar\omega = 0.048$ eV and $\lambda = 2.10$ eV.

5.4 A role for low-frequency, environmental modes

Jortner (1976) concluded that the low-frequency modes of vibration in the 'lattice' (the solvent or medium, the protein or membrane) surrounding the redox centres must not be appreciably coupled to the electron transfer in *C. vinosum* because the great range of temperature over which the rate is independent of temperature indicated a characteristic frequency $\omega/2\pi c = 400\ \mathrm{cm}^{-1}$ ($\hbar\omega = 0.050$ eV) which corresponds to an ordinary intramolecular bond vibration. However, several groups of workers find that there can be appreciable coupling to low-frequency modes in the surroundings without causing a temperature dependence in the 0–100 K temperature region.

There are three ways in which the single-mode equations (5.19) and (5.67) can show temperature independence. (i) At very low temperatures such that the vibrations are found only in their zero-point levels. Nuclear rearrangement accompanying electron transfer is then by nuclear tunnelling. For this it is necessary that $T \ll T_c = \hbar\omega/2k_B$. The theories of Hopfield (1974) and Jortner (1976) used this to reproduce the temperature-independent part of the *C. vinosum* data.

The other two ways are also permitted by equation (5.13) or (5.70). (ii) If $\lambda = \Delta E$ (or $S = p$), $E^\ddagger$ (equation (5.7)) becomes zero and the rate becomes proportional to $T^{-\frac{1}{2}}$ at any temperature. This is a mildly negative dependence. It corresponds to the P curve of Figs. 3.1 or 5.1 crossing the R curve at the bottom of the R curve. (iii) If the temperature is high enough, the exponent, $E^\ddagger/k_B T$, becomes small even with finite $E^\ddagger$ so that the pre-exponential factor begins to dominate. The temperature, T_M, for maximum rate and zero variation with temperature, found by setting the derivative with respect to T of equation (5.70) equal to 0, is:

$$T_M = (\Delta E - \lambda)^2/2\lambda k_B = 2E^\ddagger/k_B = T_c(S - p)^2/S. \tag{5.75}$$

Increase of T below T_M causes an increase in the population of the quantum levels at the height of the cross-over between the R and P curves and thus an increase in rate because, as Fig. 5.5 shows, these levels have the best overlap and, therefore, the fastest rate. Increase of T above T_M drives more of the population to higher levels causing a thinning of the levels at the cross-over height and this accounts for the decrease in rate above T_M. Temperature independence in the case of frequency change was discussed in Section 3.2.5.

5.4.1 *According to Sarai*

If we introduce soft, environmental modes, frequency ω_s, into the picture, the preceding considerations can be applied to them. Their 'p' will be that given by equation (5.65). In the case of only two modes equation (5.66) will apply. They might have $S_s = p$, producing temperature independence according to (ii) while the high-frequency 'hard' mode has temperature independence via (i). For soft modes with wave numbers in the range 10–200 cm^{-1}, $T_c = 7$–140 K. Thus a number of combinations of values of S_s and p, including $S_s > p$ will give T_M as low as 20 K, and temperature independence via (iii).

When T_c for the soft modes is much smaller than T, one can apply the high-temperature approximation to them in the manner of equations

(5.69) and (5.70). This converts equation (5.66) to:

$$W = \frac{2\pi}{\hbar}|H_{RP}|^2 \frac{\exp\{-S_k(2\bar{n}_k+1)\}}{(4\pi\lambda_s k_B T)^{\frac{1}{2}}} \sum_m \left(\frac{\bar{n}_k+1}{\bar{n}_k}\right)^{\frac{1}{2}m}$$

$$\times I_m(2S_k[\bar{n}_k(\bar{n}_k+1)]^{\frac{1}{2}}) \exp\left\{-\frac{(\lambda_s + m\hbar\omega_k - \Delta E)^2}{4\lambda_s k_B T}\right\}, \quad (5.76)$$

where $\lambda_s = S_s\hbar\omega_s$ and the hard mode is not approximated.† This is the same as Jortner's (1976) equation (3*d*), corrected. It is now actually independent of ω_s.

Sarai (1979, 1980) used essentially the foregoing considerations to advocate inclusion of soft modes with the intramolecular to explain the biological data. He argued that inclusion of soft modes not only made it easier to balance the energy changes without resort to fractional values of *p*, but also that it is difficult to conceive of soft modes not being excited by electron transfer. He suggests, however, that the protein environment may reduce and control the coupling of the soft modes for optimum effect.

He uses equation (5.66) to investigate the energy-gap dependence and temperature dependence of systems with one soft mode plus one hard mode. He calculates and presents the results of varying the frequencies and couplings of the two modes, from which several conclusions may be mentioned: (*a*) The energy-gap dependence can be periodic, as shown in Fig. 5.6, even more so at low temperatures, reflecting the vibrational structure of the hard mode. (See also Ulstrup & Jortner, 1975.) (*b*) If ΔE is slightly smaller than $m\hbar\omega_k$, that is, one of the possible hard-mode values, then *p* is required to be negative and energy is extracted from the soft modes. This requirement forces the rate to zero at temperatures low

† If the hard mode is also approximated in the high-temperature limit ($k_B T \gg \hbar\omega_k$) one obtains:

$$W = 2\pi\omega_k|H_{RP}|^2(4\pi(\lambda_k\lambda_s)^{\frac{1}{2}}k_B T)^{-1}\sum_m \exp\left\{-\frac{(m\hbar\omega_k - \lambda_k)^2}{4\lambda_k k_B T} - \frac{(\Delta E - m\hbar\omega_k - \lambda_s)^2}{4\lambda_s k_B T}\right\}, \quad (5.77)$$

where $\lambda_k = S_k\hbar\omega_k$. A plot of ln W v. ΔE for this equation will look very similar to Fig. 5.6. Note that $(2\lambda_s k_B T)^{\frac{1}{2}}$ is the standard deviation of the Gaussian functions about each peak. Jortner (1976) gives as a high-temperature limit (his equation (3e), corrected):

$$W = \frac{2\pi}{\hbar}|H_{RP}|^2(4\pi(\lambda_k+\lambda_s)k_B T)^{-1}\exp\left\{-\frac{(\Delta E - \lambda_k - \lambda_s)^2}{4(\lambda_k+\lambda_s)k_B T}\right\}, \quad (5.78)$$

which is a simple Gaussian function of ΔE. Dr Otto Sankey pointed out to me that both equations are valid, and that equation (5.77) applies when $\hbar\omega_k \gg (4\lambda_s k_B T)^{\frac{1}{2}}$ and (5.78) when $\hbar\omega_k \ll (4\lambda_s k_B T)^{\frac{1}{2}}$. Of course, sufficiently high temperature will always produce the latter condition and the oscillations of Fig. 5.6 will be smoothed out.

enough that the soft modes do not have the energy to give. On the other hand, if ΔE is slightly larger than $m\hbar\omega_k$ the soft modes receive energy and the rate approaches a constant as the temperature decreases to absolute zero.

5.4.2 *Implications for control of electron-transport rate*

Fig. 5.6 is a plot of $\log[W\hbar/2\pi|H_{RP}|^2]$ v. ΔE according to equation (5.76) for room temperature with $\hbar\omega_k = 0.05$ eV, $S_k = 4$, λ_s 0.001 eV. It suggests that kinetic 'control', C, as defined by equation (4.3) can be much greater than $\frac{1}{2}$ even with ΔE substantially positive if the system be 'engineered' for some critical value of $\Delta E - m\hbar\omega_k$. For a comparison the value $C = \frac{1}{2}$ is indicated by the slope of a dashed line while the slope for ΔE between 0.135 and 0.140 eV (e.g.) gives $C = 5.5$.

Fig. 5.6. Energy-gap law with two frequencies. Calculated according to equation (5.76), with the following values of parameters: $\hbar\omega_k = 0.05$ eV; $S_k = 4$ ($\lambda_k = 0.20$ eV); $\lambda_s = 0.00125$ eV; $4k_B T = 0.103$ eV ($T = 300$ K). The curve is practically a replotting of Sarai (1979, Fig. 1(a); 1980, Fig. 4). The slopes, C, as defined by equation (4.3) are $d(\ln W)/d(\Delta E/k_B T)$ and are a measure of the control of the rate achievable by variation of ΔE. The points, ○, are the values at $\Delta E = m\hbar\omega_k$. The peaks occur λ_s to the right of these points.

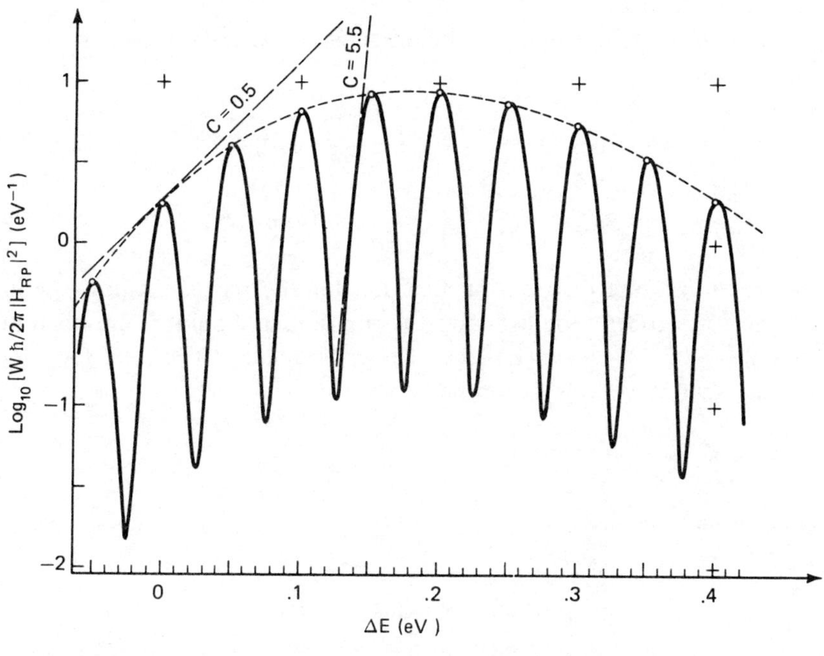

5.4.3 *According to Kuznetsov, Søndergård & Ulstrup*

In an earlier paper Kuznetsov *et al.* (1978) (Dogonadze *et al.*, 1979; Ulstrup, 1979) propose that a dispersion of the soft-mode frequencies can produce a temperature-independent rate. The argument is that, as the temperature rises, more and more of the modes change from 'quantum' (i.e. $T_c \gg T$) to 'classical', ($T_c \ll T$) and begin to show activated, Arrhenius, behaviour. With a particular distribution of frequencies, the Debye distribution, the apparent activation energy $E^{\ddagger}$, grows in proportion to T so that the ratio $E^{\ddagger}/T$ remains constant and so also $\exp(-E^{\ddagger}/k_B T)$. By equation (3.2) this produces a temperature-independent rate. The Debye distribution is reasonable as it predicts many solid state properties, such as heat capacity, quite accurately. Since this appears to be an important advance in the analysis of biological electron-transport mechanism, I think it worth while to present a brief derivation of the equations.

They start out with a formulation like equation (5.55) but containing a factor of the form of the exponential there for each vibrational mode. For the continuous distribution this is done by forming an integral of the exponents against frequency, weighted by the Debye frequency dispersion formula. Their equations use $-\theta\beta\hbar$ in place of our *it* and they proceed directly to integration over the dummy, θ, without going through the stage of equation (5.58). Their method is neat, but for the sake of uniformity in this book, I will parallel their derivation starting with equation (5.58).

In order to include one hard mode, subscript k, and a dispersion of soft modes, set:

$$G_+(t) = S_k(\bar{n}_k + 1)\,e^{i\omega_k t} + \frac{1}{\hbar}\int_{-\infty}^{\infty} \frac{d\omega}{\omega}\lambda(\omega)(\bar{n}(\omega) + 1)\,e^{i\omega t}, \tag{5.79}$$

$$G_-(t) = S_k\bar{n}_k\,e^{-i\omega_k t} + \frac{1}{\hbar}\int_{-\infty}^{\infty} \frac{d\omega}{\omega}\lambda(\omega)\bar{n}(\omega)\,e^{-i\omega t}, \tag{5.80}$$

where $\lambda(\omega)$ is the distribution function for the reorganization energies of the soft modes. For a Debye distribution it should be proportional to $(\Omega_D{}^2 + \omega^2)^{-1}$, where Ω_D is the 'Debye frequency'. The total soft-mode reorganization energy should be:

$$\lambda_s = \int_0^{\infty} \lambda(\omega)\,d\omega. \tag{5.81}$$

This is achieved if

$$\lambda(\omega) = \frac{2}{\pi}\frac{\Omega_D\lambda_s}{(\Omega_D{}^2 + \omega^2)} = \frac{2S_s}{\pi}\frac{\hbar\Omega_D{}^2}{(\Omega_D{}^2 + \omega^2)}, \tag{5.82}$$

where S_s is defined as $\lambda_s/\hbar\Omega_D$ in lieu of some simple expression for an average $\lambda(\omega)/\hbar\omega$. The quantity $2S_s\Omega_D{}^2/\pi$ will be abbreviated D.

Substituting these into equation (5.58) and working through the procedures of equations (5.59)–(5.65) at least with respect to the hard mode leads to:

$$W = \frac{|H_{RP}|^2}{\hbar^2} \exp\left\{ - S_k(2\bar{n}_k + 1) - D \int_{-\infty}^{\infty} \frac{d\omega(2\bar{n}(\omega) + 1)}{\omega(\Omega_D{}^2 + \omega^2)} \right\} \times \sum_m \left(\frac{\bar{n}_k + 1}{\bar{n}_k} \right)^{\frac{1}{2}m} I_m(2S_k[\bar{n}_k(\bar{n}_k + 1)]^{\frac{1}{2}}) \int_{-\infty}^{\infty} dt \exp[\Phi_m(t)], \tag{5.83}$$

where

$$\Phi_m(t) = \left(m\omega_k - \frac{\Delta E}{\hbar} \right) it + D \int_{-\infty}^{\infty} \frac{d\omega[(\bar{n}(\omega) + 1)e^{i\omega t} + \bar{n}(\omega)e^{-i\omega t}]}{\omega(\Omega_D{}^2 + \omega^2)}. \tag{5.84}$$

The integral over dt is evaluated approximately, by the method of 'steepest descent', a 'saddle-point' method (see, for example, Mathews & Walker, 1965). The integrand can be shown to be 'analytic'. The integral is a line integral following an arbitrary path in the complex t-plane between the specified limits. The path is chosen to go over a saddle point of the real part of the exponent of the integrand in the direction of steepest rise and descent over the saddle, which is also the direction of the contours of constant value for the imaginary part. At the saddle point Φ is expanded in a Taylor's series to the second order, giving the integrand, $\exp(\Phi_m)$, the form of a Gaussian, as follows.

First one forms the first and second derivatives of $\Phi_m(t)$ with respect to t:

$$\Phi_m{}'(t) = i\left(m\omega_k - \frac{\Delta E}{\hbar} \right) + iD \int_{-\infty}^{\infty} d\omega \frac{(\bar{n}(\omega) + 1)e^{i\omega t} - \bar{n}(\omega)e^{-i\omega t}}{\Omega_D{}^2 + \omega^2}, \tag{5.85}$$

$$\Phi''(t) = -D \int_{-\infty}^{\infty} d\omega\, \omega \frac{(\bar{n}(\omega) + 1)e^{i\omega t} + \bar{n}(\omega)e^{-i\omega t}}{\Omega_D{}^2 + \omega^2}. \tag{5.86}$$

Next one sets equation (5.85) equal to zero and solves for t^* the 'saddle point'. Thus $\Phi_m{}'(t^*) = 0$. The Taylor expansion of $\Phi(t)$ is

$$\Phi(t) \approx \Phi(t^*) + \tfrac{1}{2}\Phi''(t^*)(t - t^*)^2. \tag{5.87}$$

The integral of the exponential of 5.87 is:

$$\int_{-\infty}^{\infty} dt \exp[\Phi_m(t)] \approx \left(\frac{2\pi}{|\Phi_m{}''(t^*)|} \right)^{\frac{1}{2}} \exp[\Phi_m(t^*)]\, e^{i\phi}, \tag{5.88}$$

where $\phi = -\frac{1}{2}\arg[\Phi_m''(t^*)] \pm n/2$. The ambiguity of the sign must be resolved by taking a closer look at a plot of Re $\Phi_m(t)$ over the t-plane to determine which is the forward direction over the pass. Substitution of (5.88) into equation (5.83) gives a complete equation for the composite system:

$$W = \frac{|H_{RP}|^2}{\hbar^2}\exp[-S_k(2\bar{n}_k + 1) - \Phi(0)]\sum_m \left(\frac{\bar{n}_k + 1}{\bar{n}_k}\right)^{\frac{1}{2}m}$$
$$\times I_m\{2S_k[\bar{n}_k(\bar{n}_k + 1)]^{\frac{1}{2}}\}\left(\frac{2\pi}{|\Phi_m''(t^*)|}\right)^{\frac{1}{2}} \exp[\Phi_m(t^*)]\exp(i\phi). \tag{5.89}$$

The subscript m is attached to Φ when it depends on m but omitted from $\Phi(0)$ and $\Phi''(t)$ because they do not. $\Phi_m''(t^*)$ depends on m through t^*. Equation (5.89) is not simple as the integrals over dω must be done numerically. Kuznetsov *et al.* (1978) (Ulstrup, 1979; Dogonadze *et al.*, 1979) show calculations of a number of cases with soft modes only, demonstrating that they can fit the *C. vinosum* data below 100 K with $\lambda_s = 0.9$ eV assuming ΔE to be 0.49 eV and $\Omega_D = 150$ cm^{-1} ($\hbar\Omega_D = 0.019$ eV) or with $\lambda_s = 0.25$ eV and the same Ω_D if $\Delta E = 0.1$ eV. They do not calculate any curves with the high-frequency mode included but suggest that the *C. vinosum* data can be fitted with $\omega_k = 400$ cm^{-1} ($\hbar\omega_k = 0.050$ eV) and S_k determined by taking $E^\ddagger$ equal to 0.14 eV (or, better, 0.18 eV, see Table 4.2) and using equation (5.7) to calculate the total λ. One would then set S_k equal to (λ (total) $- \lambda_s)/\hbar\omega_k$. However, I think one should warn that according to Fig. 5.5 the apparent activation energy of 0.18 eV measured at room temperature assuming equation (3.2) does not represent a fully activated system and equation (3.2) may not be accurate at room temperature.

5.4.4 *According to Grigorov & Chernavskii*

The earliest attempt to fit the *C. vinosum* data with a theory which included vibronic coupling was made by Grigorov & Chernavskii (1972). They discussed the factors involved and, in order to represent the main features in a rough way, they picked out two pathways, one representing the non-activated nuclear tunnelling mechanism and the other a mechanism which gives an apparent activation energy and includes changes of vibrational frequency caused by the electron transfer. Translated into the symbols of this book, their equation is:

$$W = \frac{2\pi}{\hbar}|H_{RP}|^2\frac{1}{\Gamma}[4S_1S_2 + e^{-E^\ddagger/k_BT}], \tag{5.90}$$

where Γ is taken to be the width of the excited vibrational energy levels resulting from the application of the uncertainty principle to the brevity of their lifetime. It comes from using a Lorentzian instead of a δ-function to control conservation of energy. S_1 and S_2 are the coupling parameters, S, of two oscillators assumed to change quantum numbers simultaneously during the transition. They were assumed to have high frequencies so that at temperatures of interest they are initially mostly in their $n = 0$ states. They furnish the temperature-independent part of the reaction. The other term, with activation energy $E^{\ddagger}$, was conceived as arising from coupling the electron transfer to oscillators which did not change vibrational quantum number. It balances the electronic energy loss, E, with the change of vibrational energy resulting from change of frequency with no change in quantum number. The factor $|H_{RP}|^2$ was actually given as $V^2 \exp[-2b(2mV/\hbar)^{\frac{1}{2}}]$ in an effort to include its dependence on electron-transfer distance via the Gamow factor (equation (2.1)).

If one regards the parameters as purely empirical without theoretically justifying them, the model is, in effect, the same as having two parallel reaction paths, one temperature-independent and one temperature-dependent which is the way DeVault & Chance (1966) were content at that time to leave the matter. However, the paper was the first to introduce a vibronic coupling mechanism and spurred the later efforts anticipating many of their features including the role of frequency change discussed in the next section. Equation (5.90) is plotted in Fig. 5.2 with parameter values which give the best fit to the *C. vinosum* data. While it seems to fit a little better than the other two theories, the differences between all three are probably within the experimental error.

5.5 Some miscellaneous aspects

5.5.1 *Frequency changes*

The generating function method of treating electron transfer coupled to multiphonon processes outlined in equations (5.38)–(5.57) can be written so as to include changes of frequency as well as shifts of equilibrium position of the oscillators (Kubo & Toyozawa, 1955; see also Appendix, Section A.6). However, then the manipulations corresponding to equations (5.58–5.63) for integrating over t become complicated and most authors have chosen to neglect frequency changes on the basis that they are small compared to equilibrium position shifts. Lin (1966) attempted a first-order approximation for small frequency change good, however, only for the case that position shifts are equally small. He also used the non-adiabaticity operator, discussed in the next section.

Kakitani & Kakitani (1981) adapted Lin's formulation to use the exchange matrix element, $|H_{RP}|$, instead of the non-adiabaticity operator.

Their mechanism differs from that of Grigorov & Chernavskii (1972) discussed in the last section in that the change in energy content of the oscillators which change frequency is also balanced by changes in oscillators coupled to electron transfer by equilibrium point shift.

They arrive at an equation identical to equation (5.67) except that p is now equal to $-\Delta G^\circ/\hbar\langle\omega\rangle$, instead of $\Delta E/\hbar\langle\omega\rangle$, where $-\Delta G^\circ$ is the standard free energy decrease of the reaction and $\langle\omega\rangle$ appears in place of ω. Thus the energy gap is replaced by the free-energy gap as in equation (3.3) compared to (5.7). (The Kakitanis use the opposite sign for ΔG to accord with ΔE, but I am retaining the thermodynamic usage as in Marcus' theory.) $\langle\omega\rangle$ is an average (radian) frequency weighted according to the degree of coupling to the electron transfer by the coordinate shift mechanism.

To first-order approximation $-\Delta G^\circ = \Delta E - \sum_j(\bar{n}_j + \frac{1}{2})\hbar\Delta\omega_j$, where $\Delta\omega_j$ is the radian frequency change for mode j and $\bar{n}_j$ is the thermally averaged vibrational quantum number for mode j defined above equation (5.57). Thus p becomes temperature-dependent through the temperature dependence of ΔG°. To estimate this dependence they define ΔE_i as $\sum_j \hbar\Delta\omega_j$ and $\bar{n}_a$ as the average of the $\bar{n}_j$ weighted according to $\Delta\omega_j$, so that $-\Delta G^\circ$ becomes $\Delta E - (\bar{n}_a + \frac{1}{2})\Delta E_1$. For the temperature dependence of $\bar{n}_a$ they assume $\bar{n}_a = [\exp(\hbar\omega_a/k_B T) - 1]^{-1}$ (similar to that of the $\bar{n}_j$ of which it is an average) with ω_a chosen to fit the data on temperature dependence of the rate of electron transfer.

In application to data on temperature dependence of biological electron transfer such as that in Fig. 1.1, the authors assume that $\langle\omega\rangle$ represents a high frequency, $\langle\omega\rangle/2\pi c = 1000\ \mathrm{cm}^{-1}$, while ω_a corresponds to a lower frequency, $\omega_a/2\pi c = 250$–$300\ \mathrm{cm}^{-1}$. The high value for $\langle\omega\rangle$ is supported by resonance Raman spectra observations (Kitagawa, Ozaki & Kyogoku, 1978). It means that equation (5.68) (with $\Gamma(p+1)$ substituted for $p!$) can be used in place of (5.67) up to room temperature, and now temperature dependence comes only through that of p.

Quite satisfactory fits to the *C. vinosum* data are obtained with $\hbar\langle\omega\rangle = 0.125$ eV, $S = 0.5$, $-\Delta G^\circ = 0.45$ eV at 300 K and 0.88 eV at 0 K, $p = 3.6$ at 300 K and 7.0 at 0 K, $\hbar\omega_a = 0.031$ eV, $\Delta E_1 = -0.98$ eV, and A (as in equation (5.71)) $= 2.7 \times 10^8\ \mathrm{s}^{-1}$. Similar values are used to fit other photosynthetic electron-transfer data. The high frequency assigned to the modes coupled by coordinate shift and low frequency of modes undergoing frequency shift are reminiscent of the work of Grigorov & Chernavskii (1972), who, however, in their final formulation, equation (5.90), did not consider simultaneous action of the two types of oscillator shift. The Kakitanis also attempt to explain the deuterium effect of Kihara

& McCray (1973) as due to the change of zero-point energy. See Appendix (Section A.6) for more discussion of frequency shift.

5.5.2 Non-adiabaticity operator

The non-adiabaticity operator is explained in the Appendix, Section A.2, which concerns the Born–Oppenheimer approximation. See equations (A.40) and (A.41). Briefly, since nuclear motion is generally much slower than electronic motion, one can approximate the electronic part of the wave-function of a molecular system by solving for it with nuclei fixed in position. The electronic energy eigenvalues obtained this way, when plotted as a function of the nuclear positions, form adiabatic surfaces which become potential-energy surfaces for nuclear motion. Curves 1 and 2 of Fig. 3.1 B are examples of adiabatic (i.e. non-crossing) curves. However, when the nuclei are allowed to move, the wave-functions arrived at by this approximation are no longer exactly eigenfunctions and they can change spontaneously from one to another. The matrix elements causing the changes are made from the terms neglected in the approximation and are called the 'non-adiabaticity operator'. This operator involves derivatives of both the electronic and the nuclear wave-functions with respect to nuclear coordinates. See Appendix, Section A.2.

Huang & Rhys (1950) in their now classic paper considered both radiative and non-radiative transitions. Although the electron-transfer and tunnelling reactions which are the subject of this book are radiationless transitions, the fully quantum-mechanical formulations of theory discussed so far, are all based on the model used by Huang and Rhys for radiative transitions. The essential features are those of equations (5.36)–(5.39). Equation (5.67) is practically the same as equation (4.23) of Huang and Rhys. The adaptation to non-radiative transitions was made simply by replacing the electric dipole matrix element with the exchange matrix element which we call in this book, H_{RP}, or H_{ab}. However, for non-radiative transitions, Huang and Rhys used the non-adiabaticity operator with quite different results.

Kubo (1952) and Kubo & Toyozawa (1955) included the use of the non-adiabaticity operator in their development of the generating function method for multiple frequencies and changes of frequency besides coordinate displacement. Lin (1966) expanded the non-adiabaticity operator treatment for the case of small shifts of oscillator position and small frequency changes. Freed & Jortner (1970) contributed further analysis but assumed that all 'promoting' modes of oscillation (those that generated appreciable derivatives of electronic wave-functions with respect to nuclear coordinates) had no position shift. Sarai & Kakitani (1981) removed that restriction for the case of no change of frequency and showed resulting large effects of interference between modes on the energy-gap dependence of rate.

One should note that an electronic *transition* between adiabatic states 1 and 2 of Fig. 3.1 B is not an electron *transfer*. Electron localized in donor is described by state R in Fig. 3.1 A and electron in the acceptor site by state P and the electronic energies of these states form diabatic (i.e. crossing) surfaces as shown in Fig. 3.1 A. 'Exchange' or 'resonance' or 'configuration' or 'tunnelling' interaction between these states in the region around the crossing causes the splitting into the states 1 and 2 which are hybrids of R and P, with the electron not completely localized in either donor or acceptor site.

Yamosa (1982) argues that when the interaction between R and P is so slight that electron transfer is slower than the rate of nuclear passage through the crossing zone, then the splitting into states 1 and 2, calculated while holding the nuclei fixed, does not have time to manifest. Thus he prefers diabatic electronic states, formulated to depend on nuclear coordinates, as his Born–Oppenheimer states and, assuming that the non-adiabaticity operator causes the transitions between these states, he derives the same equation as Huang & Rhys (their equation (6.11)) by the generating-function method. The result is:

$$W = \pi\bar{\omega}\exp[-S(2\bar{n}+1)]\left(\frac{\bar{n}+1}{\bar{n}}\right)^{\frac{1}{2}p} \times \{B(6m+1)I_p(z) + [A - B(4\bar{n}+2)]m^{\frac{1}{2}}[I_{p+1}(z) + I_{p-1}(z)] + Bm[I_{p+2}(z) + I_{p-2}(z)]\}, \tag{5.91}$$

where $z = 2Sm$; $m = \bar{n}(\bar{n}+1)$; $\bar{n} = [\exp(\hbar\bar{\omega}/k_B T) - 1]^{-1}$; $\bar{\omega}$ = an average oscillator frequency; $p = \Delta E/\hbar\bar{\omega}$; A (= Yamosa's $C^2/\Delta W_0{}^2$) is a dimensionless constant representing the square of the matrix element given by equation (A.43) summed over the vibratory modes; B(= Yamosa's $|B|^2/2\Delta W_0{}^2$) is a dimensionless constant representing the square of a sum of such matrix elements multiplied by the position shifts of the oscillators; and S has its usual meaning in this book of a dimensionless sum of the position displacements. Yamosa obtains good fits to the *C. vinosum* data (Fig. 1.1) with, for example, $\hbar\bar{\omega} = 0.05$ eV, $p = 9$, $S = 44$, $A = 0.0012$ and $B = 0.00030$. Preliminary calculations by DeVault & Sarai (1983) indicate that the exchange interaction, neglected in forming diabatic states, may be at least equally strong in promoting the electron transfer.

5.5.3 *The reorganization energy, λ*

The values of λ required to fit the data on cytochrome oxidation in *C. vinosum* (2.24 eV for Hopfield's theory, equation (5.19), and 2.10 eV for the Jortner theory, equation (5.67) by these simple single vibrational mode theories is criticized by Warshel (1980) as too large. Warshel & Weiss (1978) calculated theoretical *S*-values for electron transfer to or from porphyrins which always came out less than 1.

Lin & Bersohn (1968) studied the effect on triplet-state lifetimes of deuteration of naphthalene and acetophenone. Deuteration slowed the radiationless decay in proportion to the number of deuterons introduced regardless of position in the molecule. They conclude that C—H stretching vibrations are important as 'accepting' modes (those having non-vanishing overlap integrals, equation (5.36)) while 'promoting' modes are more likely to be C—C stretching vibrations. The process was not electron transfer but follows most of the same quantum mechanics.

Buhks, Bixon & Jortner (1981) present a theoretical treatment of the effect on electron-transfer rates of deuterating the ligands (H_2O and NH_3) of complexes of iron and cobalt. λ_i is accounted for by the totally symmetric 'breathing' mode of vibration of the ligands and λ_0 is calculated by equation (3.5). It is also necessary to take account of the effect of deuteration on the energy gap, ΔE. See also extensive calculations by Brunschwig *et al.* (1980).

5.5.4 *Through-bond interaction*

Evidence is growing that electron interaction can occur through a series of σ-type bonds in carbon chains. These 'through-bond' interactions (TBI) were predicted by Hoffmann, Imamura & Hehre (1968) (Hoffmann, 1971) on the basis of computer calculations of molecular orbitals. Large interactions (splittings as much as 0.87 eV) have been found between donor and acceptor groups separated by 4 and 5 σ bonds by Padden-Row *et al.* (1981) and Pasman, Rob & Verhoeven (1982), in molecules so rigid that there is no chance of the interacting groups coming close enough to each other to give appreciable through-space (tunnelling) interaction. The effect is believed to arise from interaction between two π molecular orbitals and the σ molecular orbitals of the carbon chain separating the π bonds. The symmetric and the antisymmetric π orbitals formed from the two π bonds interact differently with the symmetry of the σ orbitals, causing the splitting. Symmetric and antisymmetric orbitals can be combined as in equation (A.8), Section A.1.2(*a*), to give an orbital which represents an electron jumping back and forth with a frequency equal to the energy splitting divided by *h*.

5.6 Electron tunnelling

The theory, so far, has concentrated on the nuclear motions accompanying and required by electron transfer. The electron transfer itself has been pictured simply as a crossing from an R curve to a P curve in plots of energy v. nuclear coordinates, Figs. 3.1, 5.1, 5.4 and 5.5, or as an algebraic symbol κ, in equations (3.1), (4.1) and (5.14), or $|H_{RP}|$ in equations (2.2), (5.13), (5.14), (5.19), (5.67) and (5.89) or as a non-adiabaticity operator hidden in equation (5.91). The biophysicist cannot be satisfied with this because the biological electron transfers can evidently take place over rather large distances and one is interested in the path and the dependence on the parameters of the path. The Born–Oppenheimer formalism, as generally used by the theories so far discussed, does not allow an examination of this question. It gives a picture of R–P diagrams like those of Fig. 3.1, which represent the electronic energy as determined by nuclear configuration but say nothing more about the electrons except that they take whatever configuration is most suitable to the given nuclear configuration and particular electronic 'term'.

One can argue that the electrons are indistinguishable and one cannot know that the electron leaving the donor is the same one as that arriving at the acceptor, etc. Nevertheless an electron does get from donor to acceptor and quantum mechanics can be formulated so as to discuss the problem. Van Duyne & Fischer (1974) do it in a formal manner. The following will be a more graphical presentation.

It is necessary to go back a few steps in the formulation of a Born–Oppenheimer picture. Given a certain nuclear configuration one can build up a quantum-mechanical picture of the suitable electronic structure by adding the electrons one-by-one with appropriate wave-functions formulated in the self-consistent field of all the previous electrons and before one is finished all have to be made consistent with each other, antisymmetrized, etc. (I will not try go into the specifics of molecular orbital theory). One can stop this process at the point of placing the last electron in a donor – surroundings system. This electron will be designated for computational purposes to be the one present in the donor or acceptor when either is reduced or absent when oxidized, in other words, the one that transfers. It may become lost in the antisymmetrizing process but the wave-function for the system still can be formulated with a place for such an electron.

Fig. 5.7 shows plots of the self-consistent field of all the other electrons and nuclei as experienced by the transferring electron for several nuclear configurations. The abscissa is a coordinate representing an electron moving from the vicinity of the donor to that of the acceptor plus accompanying accommodating variations in position of the other electrons, so that it actually represents transition of the whole system from the R state to the P state at fixed nuclear arrangement. The energy plotted is the total energy of the system less the kinetic energy of the electrons moving in the transition, that is, the effective potential energy for the coordinate describing the transfer. The short horizontal lines in the wells are the 'term' values for the different electronic states and match the corresponding levels plotted in the R, P diagrams (Figs. 3.1 and 5.1), that is, they represent the energy released when a formerly free electron falls into the oxidized redox centre without nuclear motion, namely the 'vertical' electron affinity of the centre. Different heights for the same centre show the influence of the different polarizations of the medium corresponding to the different nuclear coordinates. In the space between the donor and acceptor wells the plot represents the vertical electron affinity of the barrier material as influenced by the polarization of the

Fig. 5.7. Electron barriers influenced by nuclear configuration. Plots of potential energy for electronic motion involved in the change from electronic state R to electronic state P, characterized mainly by movement of an electron from donor to acceptor for several fixed nuclear configurations. The potential energy plotted is that given by the self-consistent field of all the other electrons. The labels A, B, C, R, P all correspond to the same labels in Figs. 3.1 and 5.1. The scheme used follows equation (5.107). Horizontal shading indicates potential 'wells' of the redox centres. Vertical shading indicates electron affinity of the medium.

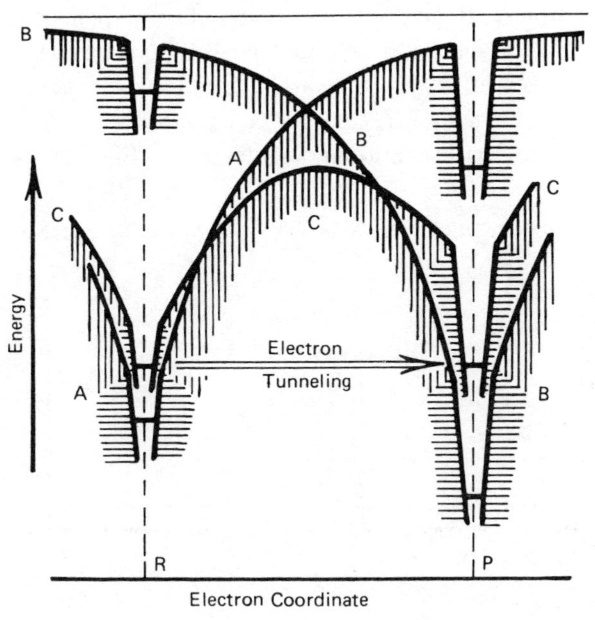

medium. It may be considered to be the height of the barrier which must be crossed by the transferring electron. The electron can tunnel the barrier when the energy levels in R and P are equal (curve C).

Whether the transferring electron 'tunnels' or not is mainly a question of whether the barrier exists. If the redox centres are more than, say, 5 Å apart and there is no metallic or conjugated-bond conduction between them, we can be quite sure that the barrier is there. If the redox centres were close enough for Van der Waals contact between them the barrier might be so small as to be negligible. Then, even if the electron tunnels the barrier, it would be expected to do it so rapidly that the process becomes adiabatic and the tunnelling is unimportant. A second criterion for (noticeable) electron tunnelling is, therefore, non-adiabaticity – as indicated by a value appreciably less than 1 for κ. This was discussed in equations (3.1) and (3.2) and in connexion with Tables 4.1, 4.2 and 4.3.

5.6.1 *One-dimensional barrier models*

If one knew the shape of the barrier one could presumably solve the appropriate Schroedinger equation and answer all questions concerning the dependence of the rate of tunnelling on the parameters of the barrier. In the absence of such knowledge a number of models have been tried in attempts to approximate one's idea of what the barrier should be like.

Equation (2.1), for a one-dimensional 'rectangular' barrier, Fig. 2.1, is perhaps the simplest. It gives P, the probability of penetrating the barrier per approach. A rate can be obtained by multiplying by a frequency of approach. However, it is desirable to relate this expression to the Hamiltonian matrix element, $|H_{ab}|$, used in equation (2.2). To do so, it is necessary to start with 'stationary' states as in Fig. 5.8. Solution

Fig. 5.8. Rectangular barrier between bound states. E_1 and E_2 are the energy eigenvalues.

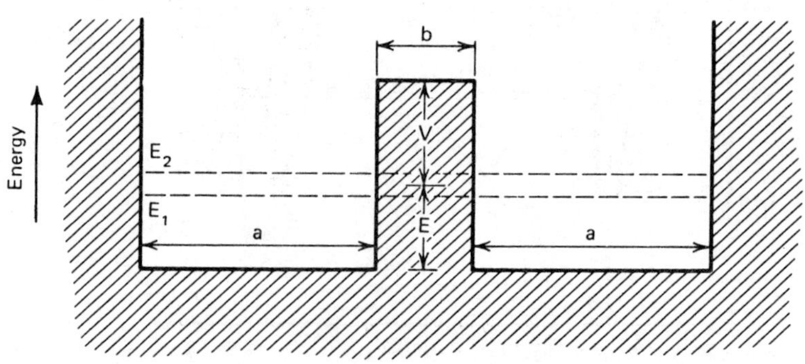

of the Schroedinger equation for Fig. 5.8 gives a pair of eigenstates with energies E_1 and E_2. The splitting between them is:

$$E_2 - E_1 = 8EV(V+E)^{-1}(1+la)^{-1}\exp(-lb), \tag{5.92}$$

where $E=(E_1+E_2)/2$; $l=(2mV)^{\frac{1}{2}}/\hbar$; and $lb \gg 1$. The time, t, required for a round trip of the particle in the box with kinetic energy E is (classically):

$$t = a(2m/E)^{\frac{1}{2}}. \tag{5.93}$$

In general, for a two-state system with equal energy in both states, the splitting between the energies of their eigenstates is (Appendix section A.1.2):

$$E_2 - E_1 = 2|H_{ab}|. \tag{5.94}$$

Comparison of equation (5.94) with (5.92) and incorporation of equations (2.1), (5.93) and (5.96) (presented below) gives:

$$|H_{ab}| = \hbar P^{\frac{1}{2}}/(\tau + t). \tag{5.95}$$

By equation (2.2) the rate is proportional to the square of this and, therefore, to the probability of penetrating the barrier, as expected. However, with $\tau \ll t$, the rate is proportional to t^{-2}, not t^{-1}. Thus t^{-1} is not a real 'frequency of approach' (which would require a wave-packet formulation) but rather it expresses the attenuation of the wave-function as it spreads over the width of the box.

Franz (1967) derived a simple expression for the time τ, required for a particle to get through a rectangular barrier once it starts, namely

$$\tau = \hbar/(EV)^{\frac{1}{2}}. \tag{5.96}$$

Thus τ is smaller, the higher the barrier, independent of b and so small that, during the moment of passage, the uncertainty in the energy must be as large as $(EV)^{\frac{1}{2}}$, which is comparable to the height of the barrier. Thus one could say that by hopping very quickly the particle hurdles the barrier under cover of the resulting cloud of uncertainty.

A more general formulation (Gamow, 1928) of equation (2.1) suitable for any shape of barrier uses the so-called 'WKB approximation' (Wentzel, 1926; Kramers, 1926; Brillouin, 1926; Dunham, 1932):

$$P = C\exp\left\{-\frac{2(2m)^{\frac{1}{2}}}{\hbar}\int_a^b [V(x)]^{\frac{1}{2}}\,\mathrm{d}x\right\}, \tag{5.97}$$

where P is the probability of penetrating the barrier, C is a proportionality constant, m is the mass of the particle, x is the coordinate of the particle through the barrier, $V(x)$ is the height of the barrier above the total energy of the particle, in other words the negative kinetic energy of the particle in the barrier, and a and b are the entry and exit points of the barrier, the points at which the kinetic energy of the particle is zero and at which it would, classically, turn around. The exponential in either equation (2.1) or (5.97) is called the 'Gamow factor'.

A bell-shaped barrier whose Schroedinger equation can be solved exactly was suggested by Eckart (1930). The symmetrical form is presented in Fig. 5.9 along with a family of solutions. One may note that the barrier affects the transmission and can cause reflexion even for energies of approach great enough to go over the top.

Several authors have favoured a parabolic shape of barrier (e.g. Bell, 1935). This is solved with equation (5.97). Christov (1960) gives an extensive discussion of it.

(*a*) *Application of Gamow factor to nuclear tunnelling and relevance to electron tunnelling.* Jortner & Ulstrup (1979) (Ulstrup, 1979; Dogonadze *et al.*, 1979) apply equation (5.97) to the nuclear barrier of Fig. 5.1. In

Fig. 5.9. The Eckart barrier. The symmetrical form is shown at the right. Its equation is $V_T(x) = V_m[\cosh(x \ln(3 + 2\sqrt{2})/w)]^{-2}$. At the left is a plot of the transmission for various particle kinetic energies, E. The same energy scale, given in the middle in units of V_m, applies to both halves of the figure. The height of the barrier above the particle energy is $V = V_T(x) - E$. The transmission probability across the barrier is given by:

$$[\cosh(\varepsilon\beta)^{\frac{1}{2}} - 1]/[\cosh(\varepsilon\beta)^{\frac{1}{2}} + \cosh(\beta - \pi^2)^{\frac{1}{2}}],$$

where $\varepsilon = E/V_m$, $\beta = 3.335\, mw^2 V_m$ = mass of particle in units of electron mass, w = width of the barrier at half-height in Å, and V_m is in electron volts. This is plotted against ε for the values of β indicated on the curves. $\beta = \infty$ gives the probability assumed by Arrhenius theory.

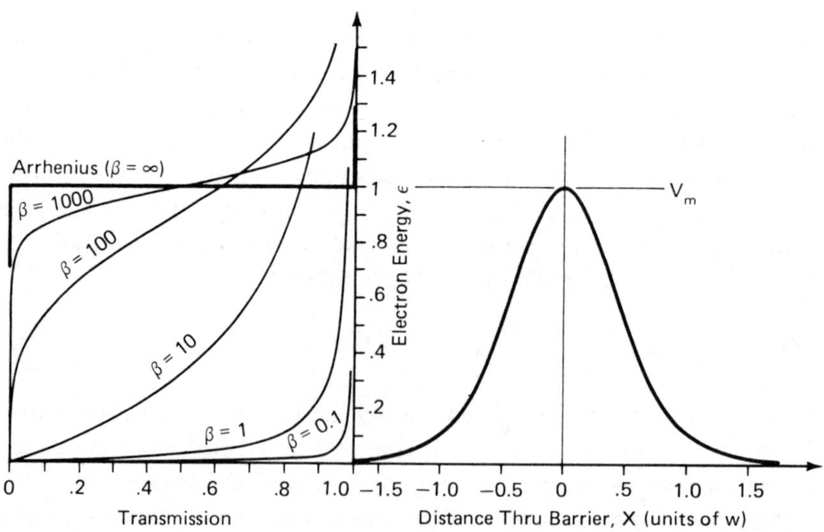

the simplest case, ignoring zero-point vibrational energy and setting $\Delta E = 0$, we may take the particle energy as equal to $E_R = E_P =$ the height of the bottoms of the parabolas, $a = X_A$, $b = X_B$ and $V(X) = V_R$ from X_A to X_C and V_P from X_C to X_B, equations (5.1) and (5.2). The height of the barrier is then $E^{\ddagger} = V(X_C)$ at it highest point. If these values are substituted into equation (5.97) the result is:

$$P = C \exp[-Q(2ME^{\ddagger})^{\frac{1}{2}}/\hbar], \tag{5.98}$$

where $Q = X_B - X_A$, the barrier 'width'. The exponent for this barrier is seen to be half of that for a rectangular barrier. The same probability of penetration is expressed by equation (5.37). For the present example, $n = p = 0$ and equation (5.37) becomes:

$$C^2(0, 0) = \exp(-S). \tag{5.99}$$

From equations (5.3), (5.4), (5.5), (5.7), and $S = \lambda/\hbar\omega$ under equation (5.20) and $\Delta E = 0$ we have:

$$S = Q(2ME^{\ddagger})^{\frac{1}{2}}/\hbar. \tag{5.100}$$

Since this is the same as the exponent in equation (5.98) it shows that the WKB approximation is not different from the calculation of Franck–Condon overlap integrals. This means also that the process envisioned by Bell (1933) with a Boltzmann distribution of particle energies, described earlier in Chapter 3, is not greatly different from that pictured in Fig. 5.5.

The same approach was applied to electron tunnelling by a number of authors, for example Weiss (1954) and Gutmann (1968). The latter attempted to estimate the temperatures below which electron tunnelling would dominate over-the-barrier reaction. However, as we have seen, the electron-tunnelling conditions depend strongly on nuclear configuration and the calculation of barrier penetration by the electron is only applicable to the activated nuclear configuration, C in Figs. 3.1, 5.1 and 5.7. The Boltzmann distribution for the kinetic energy of the tunnelling particle is more likely to be found in the nuclear motions unless there is indeed a closely spaced series of electronic terms.

5.6.2 *Three-dimensional barrier models*

One may object that electron tunnelling between redox centres is a three-dimensional problem and that one-dimensional formulas could be seriously in error. A model for the three-dimensional problem is the hydrogen molecule-ion, H_2^+ which can be solved to any desired accuracy. A simple approximate solution presented by Pauling (1928) will be described next.

A cross-section of the electron's potential energy along the axis between the nuclei is shown in Fig. 5.10. The point midway between the nuclei is a saddle point for the potential energy as it decreases in either direction from this point along the internuclear axis as shown but increases in any direction from this point perpendicular to the axis (not shown).

The resonance of the electron between the two potential wells causes a splitting of the energy into two levels, E_1 and E_2. Excluding the energy of repulsion between the nuclei, the energies of the system are given roughly by (Pauling & Wilson, 1935, p. 329):

$$E_1 = E_H + \frac{J + K}{1 + \Delta}, \tag{5.101}$$

$$E_2 = E_H + \frac{J - K}{1 - \Delta}, \tag{5.102}$$

where $E_H = -Z^2 W_H/n^2$; $J = \int \psi_a(-Ze^2/r_b)\psi_a \, d\tau$ (coulomb integral of electron cloud at a with respect to nuclear charge at b); $K = \int \psi_b(-Ze^2/r_b)\psi_a \, \mathrm{d}\tau$ (exchange integral); $\Delta = \int \psi_a \psi_b \, \mathrm{d}\tau$ (overlap integral); Z = nuclear charge in units of proton charge; $-W_H$ = energy of a hydrogen atom in its 1*s* state (13.605 eV); ψ_a = **H**-atom wave-function centred on nucleus a; ψ_b = **H**-atom wave-function centred on nucleus b; n = principal quantum number for the base states represented by ψ_a and ψ_b; e = charge of electron; r_b = distance of electron from nucleus b. E_H is the energy of an isolated H-like atom with nuclear charge Z and principal quantum number, n. The matrix element for transition from state a (electron near nucleus a) to state b (electron near nucleus b) is then

$$|H_{ab}| = \frac{E_2 - E_1}{2} = \frac{J\Delta - K}{1 - \Delta^2}. \tag{5.103}$$

Fig. 5.10. The hydrogen molecule-ion electron barrier. Cross-section along internuclear axis.

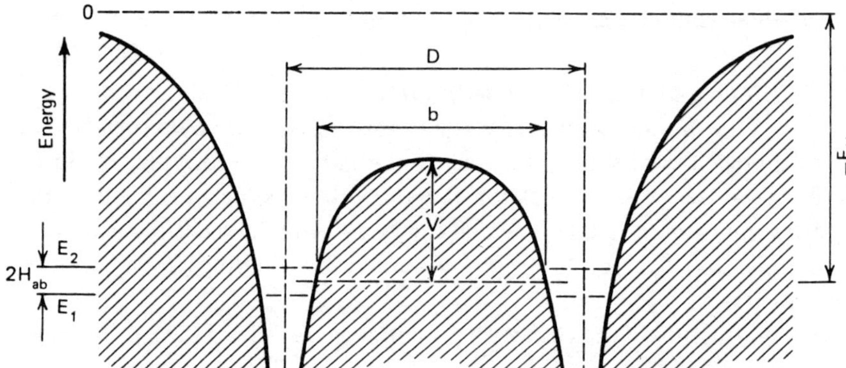

The integrals Δ and K both contain factors $\exp(-DZ/na_0)$, where D is the distance between the nuclei and a_0 is the Bohr radius, 0.529 Å. Neglecting terms with higher powers of this exponential, $|H_{ab}|$ will also be proportional to $\exp(-DZ/na_0)$. When D is very large it can approximate the barrier width, b, and the top of the barrier approaches the level 0 (see Fig. 5.10) so that the barrier height, V, approaches

$$-E \approx -E_H = Z^2 W_H/n^2.$$

Substituting these and the relation $W_H = \hbar^2/2ma_0{}^2$ into the exponent gives:

$$|H_{ab}| \propto \exp\{-b(2mV)^{\frac{1}{2}}/\hbar\}. \tag{5.104}$$

Fig. 5.11. $|H_{ab}|$ for $H_2{}^+$ and for a one-dimensional rectangular barrier. (1s) $H_2{}^+$ with base states assumed to be 1s states of the hydrogen atom. (3d): base states assumed to be $3d_{z^2}$ states of the hydrogen atom with Z-axis through the nuclei. $|H_{ab}|$ calculated according to equation (5.103). The abscissa is the internuclear distance, D. Dashed lines: $|H_{ab}|$ for a one-dimensional rectangular barrier between bound states (Fig. 5.8) calculated by equation (5.95) with $E = V$ and neglect of t. The value of V is indicated on each curve. The abscissa is the barrier width, b. The ionization potential of the hydrogen atom in its 1s state is 13.61 eV, and in the 3d state 1.51 eV.

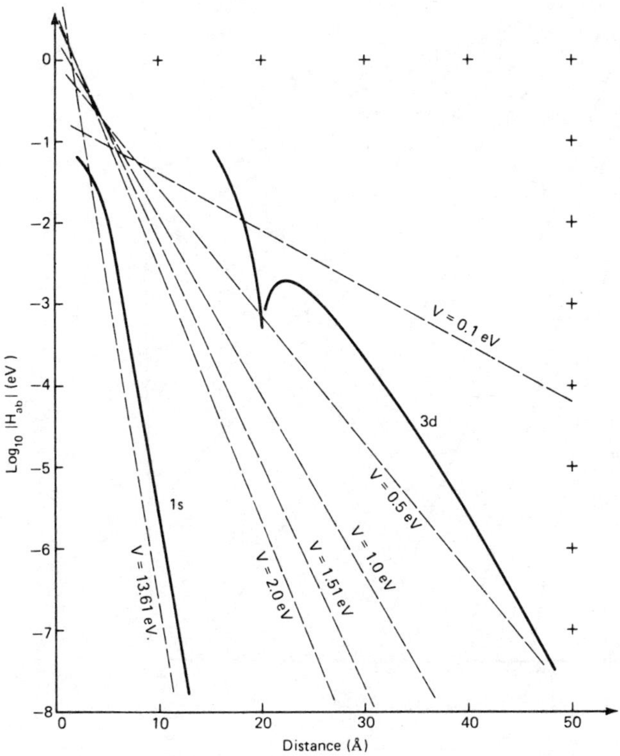

Thus $|H_{ab}|^2$ will be proportional to the same Gamow factor as given in equation (5.95) or (2.1), at least at very large distances such that the barrier shape approximates a rectangle. Going to three dimensions has not changed the Gamow factor.

Fig. 5.11 is a plot of the logarithm of $|H_{ab}|$ as a function of distance, D, between the nuclei for H_2^+ with singly charged nuclei. The labels '1*s*' and '$3d_{Z^2}$' correspond to the base-states assumed, $n = 1$ and 3, respectively. In the latter case the *d*-orbital, usually called 'Z^2' (Pauling & Wilson, 1935), was calculated with the *Z*-axis coinciding with the internuclear axis according to formulas for the integrals, J, K and Δ, given by Libby (1952) and Fig. 5.11 is similar to his Fig. 2. These calculations might be taken as approximations to the situation between two actual redox centres where

Fig. 5.12. Tunnelling in H_2^+-like systems. Nuclear charge and nuclear separation adjusted to give specified barrier heights, V, and widths, b. The calculations follow equation (5.103).

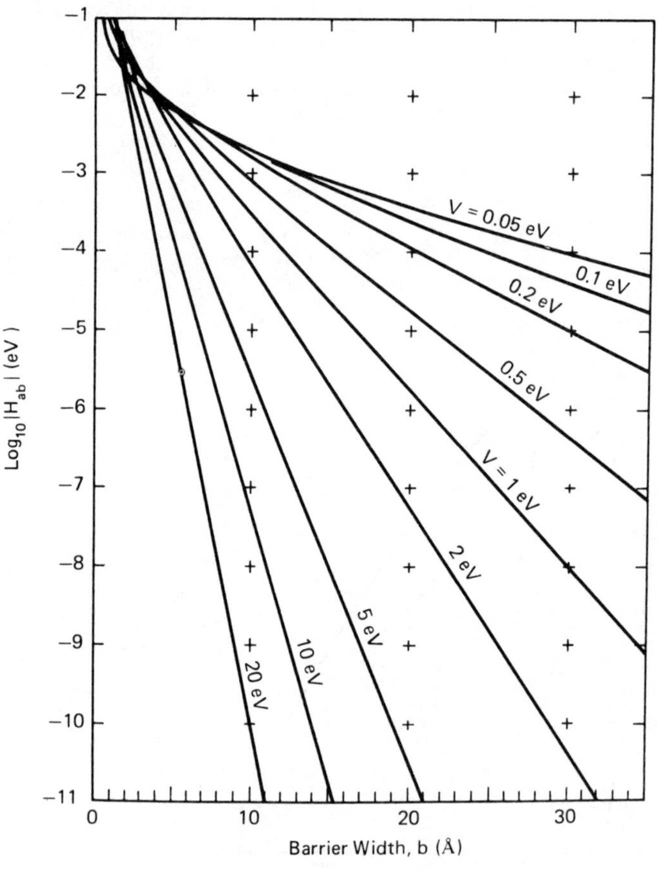

the charge on their oxidized forms is 1 +, except that the space between the centres is not likely to be empty vacuum as assumed in Fig. 5.11. Also plotted in Fig. 5.11 (dashed), for comparison, is the one-dimensional square barrier case calculated from the square root of equation (5.95) with $E = V =$ several different values. In these cases the abscissa represents the barrier width, b. In the H_2^+ case based on 1s states b is smaller than the distance between centres, D by about 2.1 Å. In the case of the $3d_{Z^2}$ basis, b is smaller than D by about 19.0 Å. If their abscissas are corrected to b-values rather than D-values the two H_2^+ curves approach closely the equation (5.95) values with $V = 13.6$ and 1.51 eV, respectively in their long-distance asymptotes, these being the ionization energies of their base-states.

One can also vary the nuclear charge in the H_2^+ model so as to produce any barrier height at any barrier width. Fig. 5.12 shows the results of such calculations using 1s base-states. Table 5.2 gives the results of such calculations using $3d_{Z^2}$ base-states and the formulas of Libby (1952).† To establish the range of interest for biological problems note that the Hopfield and the Jortner fits to the *C. vinosum* data estimate values for $|H_{ab}|$ in the range $10^{-2} - 10^{-3}$ eV.

5.6.3 *Specific expressions*

(*a*) *Hopfield's.* Hopfield (1974) used the expression:

$$|H_{RP}| = (2.7/\sqrt{20^2})\exp(-0.72b), \tag{5.105}$$

where $|H_{RP}|$ is in electron volts and b in Ångstrom units. This expression deserves further discussion as there is a tendency for other workers to adopt it uncritically to estimate tunnelling distances.

The exponential part of this expression, if squared for probability, is the same as that in equation (2.1) for a one-dimensional rectangular barrier if the height of the barrier is 2 eV. Since Hopfield states that his expression corresponds to a barrier height of 2 eV, it appears that he used the same Gamow factor.

He obtained the factor $\sqrt{20^2}$ in the pre-exponential part of the expression by consideration of the dilution of electronic contact between two porphyrins with the electron spread over say 20 atoms. The factor 2.7 eV comes from comparison with the carbon—carbon bond energy at the carbon—carbon bond distance. These choices and the choice of 2 eV barrier height are subject to several obvious criticisms, but were probably not intended to be accurate.

† Computer calculations assisted by Mr Satya Narra.

Table 5.2. *Matrix elements, H_{ab}, for tunnelling between $3d_{Z^2}$ states in H_2^+*

Barrier height, V (eV)	Barrier width, b					
	5 Å	10 Å	20 Å	30 Å	40 Å	50 Å
0.5	5.97×10^{-5} (3.97, 9.8)	6.83×10^{-6} (2.17, 18.7)	1.78×10^{-7} (1.36, 34.0)	5.54×10^{-9} (1.11, 47.2)	1.81×10^{-10} (0.98, 59.5)	6.05×10^{-12} (0.90, 71.1)
1.0	4.52×10^{-5} (4.12, 9.6)	2.89×10^{-6} (2.40, 17.9)	2.00×10^{-8} (1.61, 31.9)	1.59×10^{-10} (1.35, 44.1)	1.30×10^{-12} (1.23, 55.5)	1.07×10^{-14} (1.15, 66.6)
2.0	2.73×10^{-5} (4.38, 9.4)	7.12×10^{-7} (2.73, 17.0)	7.25×10^{-10} (1.96, 29.7)	8.11×10^{-13} (1.70, 14.2)	9.09×10^{-16} (1.57, 52.1)	9.97×10^{-19} (1.49, 62.8)

(Values in electron volts. In parentheses under each H_{ab} value are the nuclear charge (in proton units) and internuclear distance (in Å) assumed to yield the value of V in the first column and the value of b at the top of the column. The Z-axis is the internuclear axis.)

Redi & Hopfield (1980) made a theoretical study of the dependence of the electronic-tunnelling matrix element, H_{RP}, on barrier height and barrier width using two different one-dimensional models. They stress the dependence of H_{RP} on the energy level, E, of the transferring electron in that higher E means smaller net barrier height, V (see Figs. 5.8 and 5.10). Using the Hopfield model for electron transfer, equations (5.21)–(5.24) and Fig. 5.3, they average H_{RP} over E to get the effective value of H_{RP}. In applying the results to photosynthesis they suggest that the slowness of the reversed primary reactions compared to the rates of the forward reactions is partly due to the effect on H_{RP} of having to tunnel back to the chlorophyll ground state at a lower average energy level compared to the forward electron transfers. They also are concerned with the similar effects on the charge-transfer (radiative) dipole matrix element μ_{12} and estimate ratios of μ_{12} to H_{RP}.

(*b*) *Jortner's*. Jortner (1976) suggested what amounts to the expression:

$$|H_{RP}| = 5.49 \exp(-1.3b), \tag{5.106}$$

for $|H_{RP}|$ in eV and b in Å. This is a Gamow factor corresponding to a (rectangular) barrier height of 6.4 eV. The pre-exponential factor is adjusted to give $|H_{RP}| = 10\,\text{cm}^{-1}$ (1.24 meV) at $b = 10$ Å. This expression was used by Peters, Avouris & Rentzepis (1978). While some of the above criticisms do not apply here, the experimental basis for the constants chosen is still quite slim.

5.6.4 *Barrier heights*

There seems also to be no grounds for choosing 2 eV over any other value for the barrier height. A simple theory,† used in constructing Fig. 5.7, suggests that it should be equal to the difference in 'vertical' electron affinity of the barrier material and the vertical electron affinity of the oxidized redox centres, modified by the prevailing polarization of the surroundings (nuclear configuration C in Figs. 3.1, 5.1 and 5.7) at the time of electron transfer. 'Vertical' means the process of binding an electron without any nuclear motion. The electronic polarization responds but not the slower, nuclear polarization.

With the further simplifying assumptions that the oxidized forms of donor, acceptor and medium all have zero electrical charge and that their radii (of the molecules in the case of the medium) are all equal, the theory suggests that the barrier height, V, is equal to the

† See derivation in Appendix, Section A.5.

following:

$$V(x) = V_D - V_M - E^{\ddagger} + \frac{\lambda}{1-R} - \frac{R}{1-R^2}\Big\{[\lambda - E^{\ddagger}(1-R) + R\Delta E]\frac{D}{x} + [\lambda - E^{\ddagger}(1-R) - \Delta E]\frac{D}{D-x}\Big\}, \quad (5.107)$$

where V_D and V_M are the vertical electron affinities of donor and medium, respectively; R is the ratio of donor or acceptor or medium molecule radius to D, the distance between the two centres; and x is the distance from the donor centre, along the axis between centres. λ, ΔE and $E^{\ddagger}$ are related by equation (5.7) and $\lambda = (e^2/4\pi\varepsilon_0 D)(R^{-1} - 1)(\kappa_0{}^{-1} - \kappa^{-1})$ (compare with equation (3.5)), where e is the charge on an electron, ε_0 is the permittivity of free space, κ is the low-frequency dielectric constant of the medium and κ_0 is the optical frequency dielectric constant, equal to the square of the refractive index. Equation (5.107) could be used in equation (5.97).

Bietz & Miller (1979*b*) collected data on electron affinities and list the following references: Chen & Wentworth (1975), Kampars & Neilands (1977) and Szwarc & Jagur-Grodzinski (1974). Another is Vedeneyev *et al.* (1966). However, they were interested in the 'adiabatic' electron affinities, in which the product and its surroundings settle to a new equilibrium state of polarization, for use in estimating energy gaps, ΔE. To collect the vertical electron affinities more work needs to be done.

It would seem that the measurements of electron scavenging kinetics by Beitz & Miller (1979*a*, *b*), described earlier in the section on pulse radiolysis, could yield estimates of barrier heights, although their data were not analysed that way. They obtain the following expression for the distance, $R(t)$, tunnelled as a function of the average time of waiting before tunnelling that distance:

$$R(t) = \left[\frac{1}{1.09 \times 10^{-3} \cdot M} \log_{10}\left(\frac{A_0(t)}{A(t)}\right)\right]^{\frac{1}{3}}, \quad (5.108)$$

where R is in Å, M is the acceptor concentration in moles l^{-1}, A is the absorbance due to trapped electrons at time t in the presence of acceptor and A_0 the absorbance at t in a parallel experiment without acceptor.

If one plots ln t v. $R(t)$ for a given acceptor, the slope should give the attenuation coefficient, a, with respect to distance, of the tunnelling rate, k, in accordance, with

$$t^{-1} = k(R) = k_0 e^{-aR(t)}, \quad (5.109)$$

where k_0 is a constant of proportionality. Assuming that the Franck–

Table 5.3. *Barrier heights determined from the data of Beitz & Miller (1976b) according to equations* (5.107)–(5.109)

Acceptor	Abbreviation	a (Å^{-1})	V (eV)
Tetracyanobenzene	TCNB	0.89_5	0.76
Tetranitromethane	TNM	0.93	0.80
Acenaphthylene	Acnp	0.96	0.88
Benzonitrile	BzCN	0.97	0.89
Maleic anhydride	MA	1.02	0.99
9-Methylanthracene	9MA	1.04	1.03
p-Dicyanobenzene	DCNB	1.18	1.32
Naphthalene	Nap	1.22	1.42
Trimethyl benzoquinone	M_3Q	1.25	1.48
Biphenyl	Bip	1.27	1.53
Benzoquinone	BQ	1.27	1.53
Hexafluorobenzene	C_6F_6	1.35	1.73
Pyrazine	Pyz	1.37	1.79
p-Tetrachlorobenzoquinone	pCl_4Q	1.37	1.79
2, 5 and 2, 6-Dichlorobenzoquinone	Cl_2Q	1.89	3.40
Pyridine	Pyd	1.96	3.65
Benzene	Bz	4.18	16.6

Condon factors can be included in k_0 we may identify the exponential with the Gamow factor in equations (2.1) and (5.97). Thus

$$a = 2(2mV)^{\frac{1}{2}}/\hbar, \tag{5.110}$$

where V is some sort of average barrier height for irregularly shaped barriers. Table 5.3 shows values of a, measured from replots of selected parts of the data of Beitz & Miller and corresponding values of V calculated from equation (5.110). These might apply to biological tunnelling barriers to the extent that methyltetrahydrofuran can approximate a protein–lipid combination. This is presented as suggestive only. It needs more research to establish applicability. It attempts to provide an alternative explanation for what Beitz & Miller call their 'flattening' effect (departures from theoretical curves which assumed a single value, 1 eV, for all barrier heights).

5.6.5 *Distances*

One source of confusion is the difference between barrier widths, usually designated b in this book and the centre-to-centre distance of electron transfer designated by D. These were carefully distinguished in the discussion of the H_2^+ model. See Fig. 5.10.

The following estimates of distances which are independent of measurement of electron-transfer kinetics have been quoted: up to 100 Å (= b) for insulating layer thickness in tunnelling current measurements in semiconductors (Kurtin *et al.*, 1971); 25–35 Å (= D) for electrons trapped in frozen solvent tunnelling to scavenger molecules (Beitz & Miller, 1979*a*, *b*); ⩾ 25 Å (= D) between centres of cytochrome haems and RC bacteriochlorophyll in *C. vinosum* (⩾ 15 Å (= b) between edges) (Tiede *et al.*, 1978; Dutton *et al.*, 1979); ⩾ 4–6 Å (= b) between HIPIP iron – sulphur centre and cytochrome (Cusanovich & Mizrahi, 1979); ⩾ 8.4 Å (= b) between porphyrin edges of the haems in cytochrome c and cytochrome b_5 (Salemme, 1979); 8–10 Å (= D) between several nearest neighbours in the electron transport chain from succinate to quinone in mitochondria (Salerno & Ohnishi, 1979; Salerno *et al.*, 1977); 15–20 Å (= D) between haems of cytochromes c and a, a_3 in mitochondria (Vanderkooi & Erecinska, 1975; Vanderkooi & Landesberg, 1977); > 25 Å (= D) between haems of cytochrome c and cytochrome c peroxidase (Gupta & Yonetani, 1973); and 7 Å (= b) of intervening solvent between octaethylporphyrin molecules exchanging electrons (Ballard & Mauzerall, 1979; 1980). The 6.9 Å (= D) distance between Fe ion centres exchanging electrons was estimated largely by *ab initio* quantum-mechanical calculations by Newton (Sutin, 1979) but some attempt to fit the kinetic data may have influenced the distance chosen.

Attempts to estimate distances from electron-transfer rates depend strongly on the model chosen and assumptions about barrier height. To fit the *C. vinosum* rate of transfer from cytochrome c to RC bacteriochlorophyll, particularly at low temperature, we have at one extreme the calculation by DeVault & Chance (1966) ignoring Franck–Condon factors and assuming simple barrier penetration according to equation (2.1) with an electron frequency attack rate of $10^{15}\ s^{-1}$, that the barrier width, b, could be 29 Å if the barrier height, V, is 1 eV. (The reader was supposed to choose his own preferred barrier height and make his own correction.) At the other extreme we have equation (5.105) (Hopfield, 1974), which, with estimates of $|H_{RP}| = 3.6 + 10^{-3}$ eV (Hopfield) or $2.4 + 10^{-3}$ eV (Jortner) give $b = 5.0$ or 5.6 Å, respectively.

Likhtenstein *et al.* (1982) summarize distances derived from triplet transfer and spin exchange rates and apply their conclusions to a model for primary separation of charge in photosynthesis.

5.6.6 *EPR methods: measurement of* $|H_{RP}|^2/\lambda^{\frac{1}{2}}$

Measurements of interactions between electron spins on different redox centres give information concerning distances between them in several ways: (*a*) the spin–spin relaxation can affect the power level required for saturation, (*b*) magnetic dipolar interactions cause a splitting of absorption lines inversely proportional to the third power of the distance, and (*c*) electron exchange causes a splitting of absorption lines which falls off exponentially with the distance according to the Gamow factor.

Hopfield (1976) and Okamura *et al.* (1979*a*, *b*) give a relation between the exchange splitting and the matrix element for electron transfer between centres. It is similar to the difference between E_1 and H_{aa} at large negative r given by an equation like equation (A.10) in the Appendix. Thus $J/2 = |H_{ab}|^2/(H_{bb} - H_{aa})$, where J is the exchange splitting.

Okamura *et al.* (1979*a*), taking r (see Appendix) to correspond to nuclear coordinate X_A in Fig. 5.1, point out that $H_{bb} - H_{aa}$ at X_A is equal to $\lambda - \Delta E$, and from equation (5.27) this is $(4\lambda E^\ddagger)^{\frac{1}{2}}$. Substituting into the expression for $J/2$, above, one gets $|H_{ab}|^2/\lambda^{\frac{1}{2}} = JE^{\ddagger\frac{1}{2}}$. If this is substituted for $|H_{RP}|^2/\lambda^{\frac{1}{2}}$ in equation (5.12) along with a factor $\frac{1}{2}$ for the spin selection rule one gets

$$W = (J/2\hbar)(\pi E^\ddagger/k_B T)^{\frac{1}{2}} \exp(-E^\ddagger/k_B T). \tag{5.111}$$

They studied a reaction for which every quantity in equation (5.111) could be measured. They used RCs from *R. sphaeroides* with the ubiquinone replaced by menaquinone, reduced with dithionite. A flash of light then gives the reaction:

$$C\cdot P\cdot I\cdot A^- \xrightarrow{h\nu} C\cdot P^+\cdot I^-\cdot A^- \xrightarrow{10^{-5}\,\mathrm{s}} C^+\cdot P\cdot I^-\cdot A^-.$$

At 25 °C the reaction $C^+\cdot P\cdot I^-\cdot A^- \rightarrow C^+\cdot P\cdot I\cdot A^=$ proceeds with $W = 0.25\ \mathrm{s}^{-1}$. From the measured temperature coefficient of the reaction $E^\ddagger = 0.67$ eV. If $C^+\cdot P\cdot I^-\cdot A^-$ is quickly frozen its EPR spectrum at 2 K shows a doublet with a splitting of 60 G attributed to the exchange coupling between I^- and A^-. With this value for J and the measured value of W, equation (5.111) gives $E^\ddagger = 0.604$ eV. The rather good agreement between the two values for $E^\ddagger$ shows a consistency in the theory that promises an independent source of data for electron-transfer parameters. The measurement of both J and $E^\ddagger$ allows calculation of $|H_{ab}|^2/\lambda^{\frac{1}{2}} = JE^{\ddagger\frac{1}{2}}$ with less error than by measurement of W and $E^\ddagger$ and substitution into equation (5.13).

5.6.7 *Super-exchange*

Another method of estimating, theoretically, electron-transfer matrix elements, H_{RP}, was called 'super-exchange' by Anderson (1950), who drew upon work by Kramers (1934) to explain the coupling of electron spins over distances which included non-magnetic atoms in between, and who received the Nobel prize in 1977 for this and related work in magnetism. McConnel (1961) used it to estimate $|H_{RP}|$ for transfer of an extra electron between aromatic rings separated by chains of CH_2 groups, the purpose being to correlate data on such exchanges gathered by Weissman (1958) and by Voevodskii *et al.* (1959). The method uses the LCAO (linear combination of atomic orbitals) formulation of molecular orbital theory.

The problem is that in using LCAO theory one usually simplifies by postulating interaction only between nearest neighbour atoms in a molecule and therefore rules out at the start any direct interaction, such as tunnelling between donor and acceptor with barrier atoms between them. Instead, the theory postulates 'virtual' states in the barrier atoms for the electron to occupy as it moves from donor to acceptor in steps between nearest neighbours only. These virtual states are states of the negative ion of the barrier atoms in which the extra electron is in otherwise unoccupied excited states. They are called 'virtual' because their formation requires more energy than is available to the system from its original state with the electron in the donor. Thus virtual states, like the tunnelling concept, are classically forbidden. It seems to me that super-exchange is really just an alternative method of estimating quantum-mechanically the classically forbidden electron transfer which is more suitable than direct, under the barrier tunnelling to the LCAO method. Schrieffer (1979*b*) has stated the same idea in different words. In postulating that the electron rides over the barrier through virtual states, super-exchange reminds one of Franz's (1967) analysis of tunnelling, already mentioned (see Equation (5.96)) which finds that the particle could momentarily have enough energy to go over the barrier by uncertainty principle.

McConnell thus constructed virtual molecular orbitals in the paraffin barrier chain from carbon atom 3*d*-orbitals. These then perturbed the ground states of the extra electron when in either of the aromatic rings at the ends. The amount of perturbation depended on the symmetry of the ground states so that a symmetry-dependent splitting in the ground state energies resulted. His formula for the splitting is:

$$2|H_{RP}| = -2(T^2/D)(-t/D)^{n+1}, \tag{5.112}$$

where D is the energy required by the electron to reach the virtual *d*-orbitals from the ground state in either ring, t is the matrix element for transfer from one *d*-orbital to the next, T is the matrix element for transfer from the ground state in either ring to the *d*-orbital in that ring, and n is the number of carbon atoms in the paraffin chain.

One can rewrite Equation (5.112) as:

$$|H_{RP}| = (T^2t/D^2)\exp\{-b\ln(D/t)/d\}, \tag{5.113}$$

where d is the length of hydrocarbon chain per CH_2 group, and $b = nd$ represents the barrier width. Further comparison with Equation (5.95) would require matching $\ln(D/t)/d$ with $(2mV)^{\frac{1}{2}}/\hbar$ the equality of which is not immediately apparent. One would expect D to be closely related to V.

Super-exchange has been suggested as an alternative path to tunnelling, possibly even allowing greater distances of transfer (see, for example, Jortner, 1976). I would suggest that two different quantum-mechanical methods of solving the same problem should give the same answers within the limits of their respective approximations.

Dogonadze *et al.* (1973*a*, *b*) considered the somewhat different effect of vibronic coupling on the transfer to and from virtual intermediate states and applied it to estimate the catalytic effect of bridging groups on electron-transfer reactions. They show that the bridging group can provide crossing points on an R, P diagram (such as Fig. 3.1 but with three parabolas) between initial or final states and an intermediate state such as the virtual state: 'electron on the bridge'. These crossing points can be considerably below the direct crossing point between the initial and final states, and like a tunnelling path, reduce the activation energy required for the nuclear motions.

5.7 Other approaches

Although this book cannot go into an explanation of them, I should mention some other approaches to the problem of radiationless transitions.

Gouterman (1962) suggested a theory of phonon emission or absorption patterned after the usual theory of emission or absorption of photons.

A school of theoretical investigation that puts less emphasis on the Kubo–Lax method of handling the vibronic interactions than is done in this book is represented, for example, by Robinson & Frosch (1962, 1963), Siebrand (1967), Henry & Siebrand (1973), Tric (1971), and Freed (1978). They tend to emphasize the detailed quantum mechanics of the transitions from one state to an array of states using Green's function methods, elaborating, as it were, on the golden rule and variations. A simplified discussion of this type is given in the Appendix on derivation of the golden rule.

A general review of radiationless processes and an explanation of Green's function methods is found in Jortner, Rice & Hochstrasser (1969).

A careful reconsideration of the semiconduction type of mechanism is given in considerable detail by Petrov (1979).

APPENDIX

Review of pertinent quantum mechanics

A.1 The standard two-state system

This review is intended to aid the dispersion to experimental workers in the biological field of the pertinent present knowledge of quantum mechanics which has grown up in the fields of physics and chemistry. For many readers this appendix is unnecessary but my own experience and the lessons of the Philadelphia Tunnelling Conference (November 1977), Chance *et al.* (eds.) (1979*a*), show that many of us need at least parts of it. I assume the reader has an elementary knowledge of quantum mechanics, at least as taught in general physics and chemistry courses: uncertainty principle, Schroedinger equation, etc. I suppose not all readers are familiar with transformation theory (Dirac, 1927, 1935; Jordan, 1927*a*, *b*) (how to change from one representation of a system to another) and some other concepts very pertinent to tunnelling discussion which I will try to define and explain in elementary terms. An excellent beginning text for further reading is the Feynman lectures (Feynman, Leighton & Sands, 1965). A few notions which the reader undoubtedly knows already will be repeated anyway for emphasis and completeness.

Quantum mechanics attempts to tell us what we can know about the future of a system, given what we know about it now. Since scientific knowledge is always of a statistical nature and never complete, quantum mechanics quite properly deals with *probabilities*. However, it does not deal with the probabilities directly. It deals with an underlying structure of 'probability *amplitudes*' from which probabilities can be calculated by taking the squares of the absolute values of the probability amplitudes. The 'amplitudes' themselves are, in general, complex numbers but the probabilities, of course, must be real and positive.

One describes a system in terms of a set of '*base-states*'. The set must be *complete* in that any state of the system to be contemplated must be describable as a combination of the base-states. Also the base-states must be *orthogonal* and *normalized*, that is, if the system is known to be in a single base-state it must have zero amplitude to be in any other and unit amplitude to be in itself. Different sets of base-states may be used to describe the same system. Sometimes one set or 'representation' is more convenient and sometimes another.

Fig. A.1. Tunnelling between two boxes. (A)–(C) Energy v, particle coordinate, as in Fig. 5.8. Boundaries of shaded areas are potential energy. Solid horizontal lines within boxes represent the energy eigenvalues, dashed lines the values of H_{aa} and H_{bb}. (A) Box *a* has lower energy than *b*. (B) Boxes have equal energy. (C) Box *a* has higher energy than *b*. (D), (E) and (F): Graphs of occupation of states corresponding to the potential-energy diagrams (A), (B) and (C), respectively. The occupations or probabilities are shown partitioned between base-states *a* and *b* and between eigenstates 1 and 2 according to equations (A.11). (G) A plot of the energies *v*. a continuous scale of separation between H_{aa} and H_{bb}. The abscissa is $r = (H_{aa} - H_{bb})/2|H_{ab}|$. See equations (A.9).

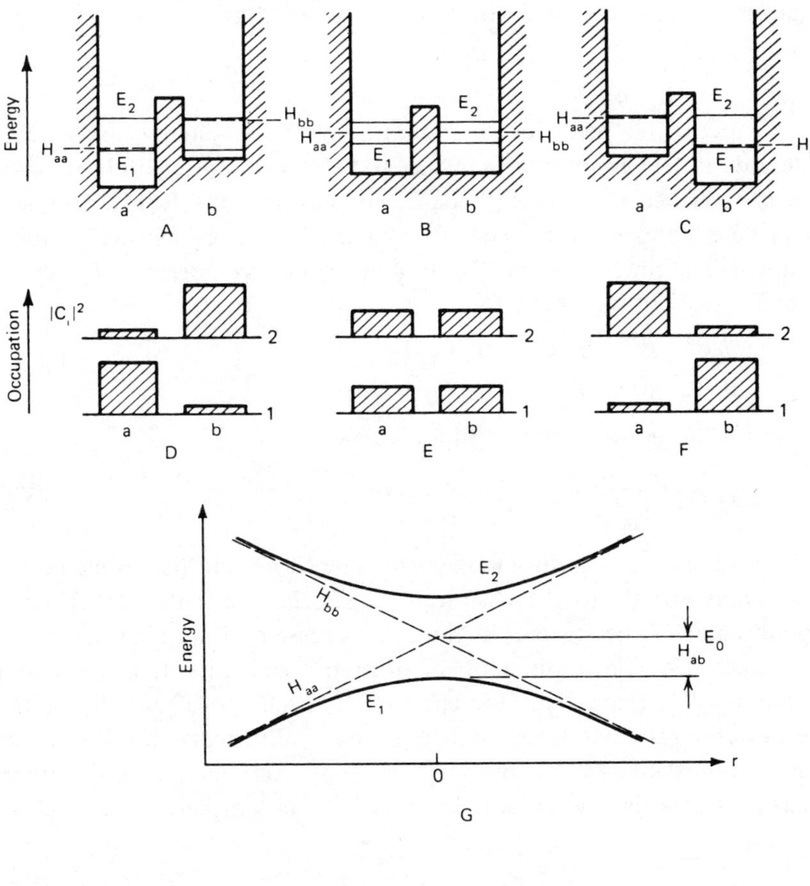

Physical quantities are represented as *operators* which affect the probability amplitudes by operating on them. Here we will be concerned mainly with the energy operator which is also called the '*Hamiltonian*'. We will discuss how this works in the case of a system which has only two base-states. This may sound like an oversimplified system but actually a large number of physical problems can be discussed in terms of the standard two-state system (Feynman *et al.*, 1965), including many of the tunnelling concepts.

Consider a system that consists of two boxes with one particle which can be in either box. There is a wall between the boxes that the particle cannot penetrate in a classical manner because it does not have enough kinetic energy. (This 'wall' is not a real one, penetration of which requires loss of energy (to heat) through inelastic collision with its material, but is, rather, a force-field which classically slows the particle on entering (to stopping if the particle does not have enough momentum to get over the hump) and speeds it on leaving.) The particle can, however, get from one side to the other by quantum-mechanical tunnelling. Fig. A.1B shows the situation on an energy v. particle-coordinate diagram.

A.1.1 *The Hamiltonian*

Let C_a be the probability amplitude for the system to be in base-state a, described as 'particle in box a', and C_b the amplitude to be in base-state b, described as 'particle in box b'. The states a and b form a complete set of base-states for this two-state system. C_a and C_b can be complex numbers. The time operator, $i\hbar \mathrm{d}/\mathrm{d}t$, and the energy operator, H, act on C_a and C_b as follows:

$$\left.\begin{aligned} i\hbar\, \mathrm{d}C_a/\mathrm{d}t &= H_{aa}C_a + H_{ab}C_b, \\ i\hbar\, \mathrm{d}C_b/\mathrm{d}t &= H_{ba}C_a + H_{bb}C_b. \end{aligned}\right\} \tag{A.1}$$

The Hs can be set forth in a square array:

$$\mathbf{H} = \begin{pmatrix} H_{aa} & H_{ab} \\ H_{ba} & H_{bb} \end{pmatrix}, \tag{A.2}$$

which is called the 'Hamiltonian matrix'. The Hs are the 'matrix elements'. They apply strictly to the base-states a and b. The matrix (A.2) is the Hamiltonian of the system 'in the a, b representation'. Another set of base-states will take a different set of matrix elements. It is seen from equations (A.1) that H_{aa} is the effect of C_a itself on $\mathrm{d}C_a/\mathrm{d}t$. H_{aa} is the energy of the system in state a if there were no influence on it from state b, that is, no possibility of the particle tunnelling from one box to the other. Likewise H_{bb} is the energy of the system in state b if there were no effect from a; H_{ab} is the effect of C_b on $\mathrm{d}C_a/\mathrm{d}t$ and H_{ba} is the effect of C_a on $\mathrm{d}C_b/\mathrm{d}t$. They represent the rate at which probability *amplitude* flows from one state to the other. If equations (A.1) were divided through by $\hbar$ they would have the appearance of chemical rate equations with first-order rate constants ($H/\hbar$), except for the presence of the imaginary root, $i(=\sqrt{-1})$. In order that the energy be real it is necessary that the diagonal elements H_{aa} and H_{bb} be real and that $H_{ab} = H^*_{ba}$, where '*' means 'complex conjugate'. Thus,

$$|H_{ab}| = |H_{ba}|.$$

A.1.2 *Eigenstates and eigenvalues*

(*a*) *Symmetrical case.* Equations (A.1) can be solved easily if the Hs do not change with time. To simplify further, let $H_{aa} = H_{bb} = E_0$. The particle has the same energy in either box. Let, also,

$$H_{ab} = H_{ba} = -A.$$

Then if we add the equations (A.1) we get:

$$i\hbar\, \mathrm{d}(C_a + C_b)/\mathrm{d}t = (E_0 - A)(C_a + C_b). \tag{A.3}$$

Subtracting the second of (A.1) from the first gives

$$i\hbar\, \mathrm{d}(C_a - C_b)/\mathrm{d}t = (E_0 + A)(C_a - C_b). \tag{A.4}$$

The form of these equations allows a simpler description of the states: suppose we let $C_1 = (C_a + C_b)/\sqrt{2}$ be the amplitude to be in a state we'll call '1' and $C_2 = (C_a - C_b)/\sqrt{2}$ be the amplitude to be in a state called '2'. (The $\sqrt{2}$ is necessary for proper normalization). Then equations (A.3) and (A.4) become

$$\left.\begin{aligned} i\hbar\, \mathrm{d}C_1/\mathrm{d}t &= E_1 C_1, \\ i\hbar\, \mathrm{d}C_2/\mathrm{d}t &= E_2 C_2, \end{aligned}\right\} \tag{A.5}$$

where $E_1 = E_0 - A$ and $E_2 = E_0 + A$.

States 1 and 2 may be taken as a new set of base-states. In this representation, equations (A.5) show that the Hamiltonian matrix is

$$\mathbf{H} = \begin{pmatrix} H_{11} & 0 \\ 0 & H_{22} \end{pmatrix}, \tag{A.6}$$

where $H_{11} = E_1$ and $H_{22} = E_2$. We have 'diagonalized' the matrix.

The equations (A.5) are seen to be independent of each other and solutions are

$$C_1(t) = C_1(0)e^{-iE_1t/\hbar} \text{ and } C_2(t) = C_2(0)e^{-iE_2t/\hbar}. \tag{A.7}$$

C_1 and C_2 oscillate in the complex plane with radian frequencies $E_1/\hbar$ and $E_2/\hbar$ respectively.

Their probabilities, $|C_1|^2$ and $|C_2|^2$, stay constant with time. States 1 and 2 are, thus, stationary states or *eigenstates*. Their energy *eigenvalues* are E_1 and E_2. The *splitting* between the two energies is seen to be $E_2 - E_1 = 2A = -2H_{ab}$. This splitting is the result of the particle's ability to tunnel from one box to the other. It is shown in Fig. A.1 B. (Note: other phase factors than -1 could have been used for H_{ab} and H_{ba} but the results would be practically the same.)

The representation is in terms of eigenstates whenever the matrix is diagonalized. On the other hand if off-diagonal elements had finite values this would indicate that the base-state amplitudes would be subject to change over a period of time and so such base-states are not stationary states.

If the particle is initially in a, how fast does it get into b and what is the future time-course of the system? This is answered by noting that from the definition of C_1 and C_2 if $C_a(0) = 0$ then $C_1(0) = C_2(0) = 1/\sqrt{2}$. Also note, from the definition, that

$$C_a = (C_1 + C_2)/\sqrt{2} \text{ and } C_b = (C_1 - C_2)/\sqrt{2}.$$

If we substitute the solutions (A.7) for C_1 and C_2 into these expressions we get

$$C_a(t) = e^{-iE_1t/\hbar} + e^{-iE_2t/\hbar} = e^{-iE_0t/\hbar}\cos(At/\hbar),$$
$$C_b(t) = e^{-iE_1t/\hbar} - e^{-iE_2t/\hbar} = i\,e^{-iE_0t/\hbar}\sin(At/\hbar),$$

$$\left.\begin{aligned} |C_a|^2 &= \cos^2(At/\hbar) = \tfrac{1}{2} + (\tfrac{1}{2})\cos(2At/\hbar),\\ |C_b|^2 &= \sin^2(At/\hbar) = \tfrac{1}{2} - (\tfrac{1}{2})\cos(2At/\hbar). \end{aligned}\right\} \tag{A.8}$$

Equations (A.8) show that the probability of finding the particle oscillates back and forth between the two boxes with a frequency equal to the splitting $(E_2 - E_1)$ divided by $\hbar$. This is the 'resonance' phenomenon (familiar to chemists) occurring in molecules with two equivalent electronic structures. The lowering of E_1 below E_0 is the 'resonance lowering' of energy and it results from the ability of the particle (electron) to go back and forth between the two structures. It is always accompanied by the raising of a companion state to a higher energy. The time-course of $|C_b|^2$ is plotted in Fig. A.2 A.

The system must be in state 1 to be in the lower energy state. How are states 1 and 2 to be described? If the system is known to be in state 1 then $C_1 = 1$ and $C_2 = 0$. From the definitions we then have

$$C_a + C_b = \sqrt{2} \text{ and } C_a - C_b = 0.$$

Thus in state 1 we have $C_a = C_b = 1/\sqrt{2}$. The particle has equal probability to be in both boxes ($\tfrac{1}{2}$ in each). If the system is in state 2 then $C_1 = 0$

and $C_2 = 1$ and $C_a + C_b = 0$ and $C_a - C_b = \sqrt{2}$. Thus in state 2: $C_a = -C_b = 1/\sqrt{2}$. Again we have equal *probability* ($\frac{1}{2}$ each) to be in both a and b but the *amplitudes* to be in a and in b are *symmetrical* in state 1 and *antisymmetrical* in state 2.

Note that one cannot observe both the going back and forth and the splitting of the energy at the same time. To observe the resonance oscillation one should make observations at intervals of time considerably less than one period, which is $h/(2A)$. The uncertainty in the time measurements must be considerably less than $h/(2A)$. By the uncertainty principle the uncertainty in energy must then be considerably larger than $2A$, which is the splitting. Thus the splitting would be unobservable. To be able to tell in which of the two energy eigenstates the system is, one would have to observe it over a period of time long enough to reduce the uncertainty in energy to considerably less than $2A$. By the uncertainty principle the observation would require a time considerably longer than $h/(2A)$ the period of the oscillation. Therefore, if one could measure the

Fig. A.2. Time-course of $|C_b|^2$, the occupancy of box b, starting with the particle in box a at time 0. (A) $H_{bb} = H_{aa}$. (B) $H_{bb} = H_{aa} \pm 2|H_{ab}|$. (C) $H_{bb} = H_{aa} \pm 4|H_{ab}|$. (D) Distributed H_{bb}; ρ = number of H_{bb} values per unit of energy.

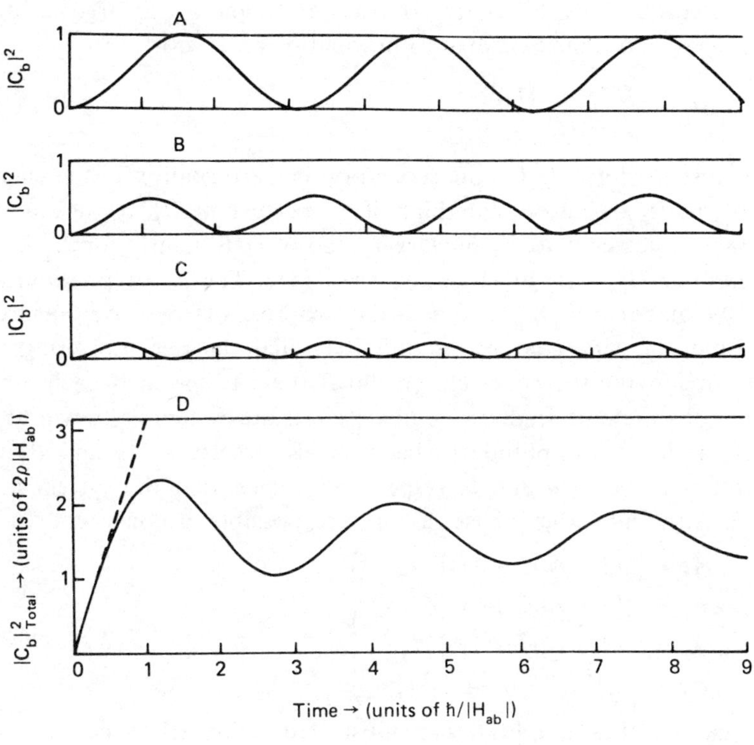

energy accurately enough to know that the system is in state 1 (or in state 2) one could observe the occupancy of the boxes only as an average over a number of periods and would find simply equal probability to be in each box.

(*b*) *Energies different*. Next, we consider the two-box case in which the particle does not have the same energy in each box. $H_{aa} \neq H_{bb}$. Let $r = (H_{aa} - H_{bb})/(2A)$ be a dimensionless representation of this energy difference where, as before, $A = -H_{ab} = -H_{ba}$, is considered to remain constant as r varies. Let $E_0 = (H_{aa} + H_{bb})/2$ be the mean of the two self-energies. For constant Hs the equations (A.1) can be solved with a little algebra and the energy eigenvalues are found to be

$$\left.\begin{aligned} E_1 &= E_0 - A(r^2+1)^{\frac{1}{2}}, \\ E_2 &= E_0 + A(r^2+1)^{\frac{1}{2}}. \end{aligned}\right\} \tag{A.9}$$

The difference, $E_2 - E_1$, is now $2A(r^2+1)^{\frac{1}{2}}$. At large values of r this approaches $H_{aa} - H_{bb}$, indicating that any resonance effect has become small and the eigenvalues are almost the same as what the energies in the individual boxes would be if separated. The behaviour of the energies is illustrated in Fig. A.1, A–C, G.

The difference between E_1 and H_{bb} or between E_2 and H_{aa} at large values of r (or between E_1 and H_{aa} or between E_2 and H_{bb} at large negative values of r) can be approximated from equation (A.9) with:

$$H_{bb} - E_1 \approx \frac{|H_{ab}|^2}{(H_{aa} - H_{bb})}. \tag{A.10}$$

This is just the formula for the second-order perturbation of the energy level H_{bb} by H_{aa}. Thus one can think of H_{aa} perturbing H_{bb} by the amount necessary to lower it to E_1 (at large value of r). Similarly for $E_2 - H_{aa}$ at large r or $H_{aa} - E_1$ at large negative r, etc. The a, b representation would be diagonal if $|H_{ab}| = 0$ as if the two boxes were moved far from each other or separated by an infinitely high barrier. Lowering the barrier or moving the boxes closer allows them to 'perturb' each other.

Let A_1 be the amplitude to be in state a if the system is known to be in state 1, B_1 the amplitude to be in b, given state 1, A_2 and B_2 the amplitudes to be in a and b, respectively, given that the system is in state 2. Although other phase factors are possible, a consistent set is:

$$\left.\begin{aligned} A_1 &= \{\tfrac{1}{2}[1 - r/(r^2+1)^{\frac{1}{2}}]\}^{\frac{1}{2}}, \\ B_1 &= \{\tfrac{1}{2}[1 + r/(r^2+1)^{\frac{1}{2}}]\}^{\frac{1}{2}}, \\ A_2 &= \{\tfrac{1}{2}[1 + r/(r^2+1)^{\frac{1}{2}}]\}^{\frac{1}{2}}, \\ B_2 &= -\{\tfrac{1}{2}[1 - r/(r^2+1)^{\frac{1}{2}}]\}^{\frac{1}{2}}. \end{aligned}\right\} \tag{A.11}$$

The squares of these amplitudes are illustrated in Fig. A.1 D–F.

If at $t = 0$ the particle is in box a, then the time-course of the probabilities to be in the two boxes become:

$$\left.\begin{aligned} |C_a|^2 &= 1 - (r^2+1)^{-1}(1-\cos^2(\omega_{21}t/2)) \\ &= 1 - \tfrac{1}{2}(r^2+1)^{-1}(1-\cos(\omega_{21}t)), \\ |C_b|^2 &= (r^2+1)^{-1}\sin^2(\omega_{21}t/2) \\ &= \tfrac{1}{2}(r^2+1)^{-1}(1-\cos(\omega_{21}t)), \end{aligned}\right\} \quad \text{(A.12)}$$

where $\omega_{21} = (E_2 - E_1)/\hbar = 2A(r^2+1)^{\frac{1}{2}}/\hbar$. As r^2 increases the particle tends to stay more and more in the box it started from, the excursions into the other box becoming smaller in amplitude and higher in frequency. Two examples are plotted in Fig. A.2 B, C.

For large r the coefficient A_1 can be approximated:

$$A_1 \approx A/(H_{aa} - H_{bb}) = H_{ab}/(H_{bb} - H_{aa}). \quad \text{(A.13)}$$

This is the perturbation theory expression for the amount of state a amplitude to be mixed into the amplitude for state b (to give the state 1) because of perturbation from state a (assuming the situation corresponding to large r).

A.1.3 *Derivation of the 'golden rule'*

The tunnelling reactions discussed in this review are not seen to oscillate as in the above examples. Nor do they show a time-course

Fig. A.3. Energy levels for an ensemble of box-pairs.

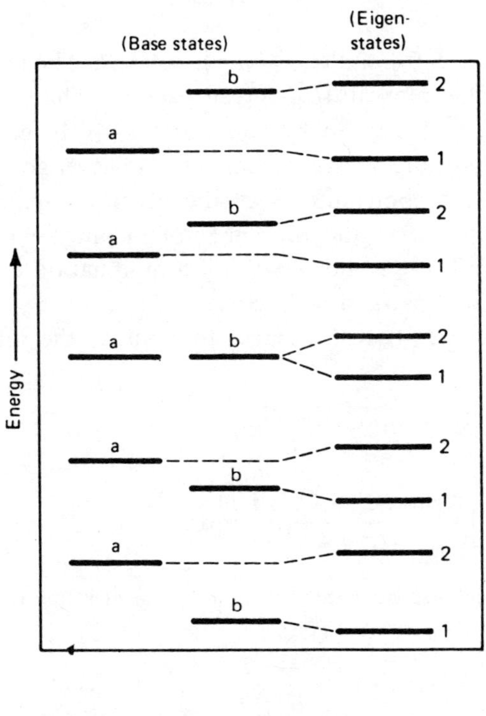

proportional to t^2 as do the cosine functions in equations (A.8) and (A.12) at small values of t. The electron tunnels to its new position and stays there. The time-course at small t is proportional to t. This is partly because its new state, state b, can sometimes decay with loss of energy to a third state from which it cannot return because of the loss of energy. However, it also results if the difference, $H_{bb} - H_{aa}$, is not a fixed single value but has a statistical distribution over a wide range of values as expected in the usual molecular situation. This requires explanation.

To simulate such a situation in an easily understood manner, imagine an ensemble of many pairs of a, b boxes (Fig. A.3). (If the ensemble of box-pairs seems strange, note that in the photosynthetic studies one observes some 10^{15} cytochrome–chlorophyll pairs all prepared in the state a simultaneously by a flash of light.) Let H_{aa} and A be the same in each box-pair, but let H_{bb} be distributed at evenly spaced values over a wide range both below and above H_{aa}. Let the spacing between neighbouring values of H_{bb} be ρ^{-1}.

At $t = 0$ the particle in each box-pair is in the a box of that pair. At a later time, t, one observes the total occupation of b boxes. Mathematically this will be a summation of $|C_b|^2$ values as given by equation (A.12). If the spacing between values of H_{bb} is made smaller and smaller and the number of box-pairs increased indefinitely we approach the integral:

$$|C_b|^2_{\text{total}} = \rho \int_{-\infty}^{\infty} (r^2 + 1)^{-1} \sin^2(\omega_{21} t/2)\, \mathrm{d}E, \tag{A.14}$$

where $E = H_{bb} - H_{aa}$, $r = -E/2A$ and $\omega_{21} = (E^2 + 4A^2)^{\frac{1}{2}}/\hbar$. The number of box-pairs included in the interval $\mathrm{d}E$ is given by $\rho\,\mathrm{d}E$. The curve of Fig. A.2 D† is a plot of $|C_b|^2_{\text{total}}$ as given by equation (A.14). In contrast to Fig. A.2 A–C, it rises in proportion to t at small t. It also never goes back to zero. This is because the contributions of the other box-pairs are oscillating at different frequencies and they are not in phase with the dominating resonant pair. There is, however, still a modulation coming from the dominance of the box-pairs with H_{bb} close to H_{aa}.

Differentiating equation (A.14) with respect to t under the integral sign gives:

$$\begin{aligned} \frac{\mathrm{d}}{\mathrm{d}t}(\Sigma C_b{}^2) &= \frac{\rho}{2} \int_{-\infty}^{\infty} \frac{\omega_{21} \sin(\omega_{21} t)}{r^2 + 1} \mathrm{d}E \\ &= \frac{4A^2\rho}{\hbar} \int_0^{\infty} \frac{\sin\left((E^2 + 4A^2)^{\frac{1}{2}} t/\hbar\right)}{(E^2 + 4A^2)^{\frac{1}{2}}} \mathrm{d}E \end{aligned}$$

† Plotted by computer approximation with help from Mr (now Dr) Robert Overfield.

$$-\frac{4A^2\rho}{\hbar}\int_1^\infty \frac{\sin(2AZt/\hbar)}{(Z^2-1)^{\frac{1}{2}}}\,\mathrm{d}Z, \qquad \text{(A.15)}$$

where $\Sigma C_b{}^2 = |C_b|_{\text{total}}{}^2$ and $Z = (E^2 + 4A^2)^{\frac{1}{2}}/2A$. The integral in equation (A.15) is $\pi/2$ times the zero-order Bessel function of $2At/\hbar$ (Watson, 1945, p. 170, eqn (3)).† Thus:

$$\frac{\mathrm{d}}{\mathrm{d}t}(\Sigma C_b{}^2) = \frac{2\pi A^2\rho}{\hbar} J_0(2At/\hbar). \qquad \text{(A.16)}$$

At very small t, since $J_0(0) = 1$, this gives a practically initial slope equal to:

$$W_{ab} = (2\pi/\hbar)|H_{ab}|^2\rho, \qquad \text{(A.17)}$$

where A was replaced by $|H_{ab}|$ and $\mathrm{d}(\Sigma C_b{}^2)/\mathrm{d}t$ by W_{ab}, the rate of going over from a to b. The form of J_0 assures that the oscillations seen in Fig. A.2 D will damp out. Equation (A.17) is the same as equation (2.2).

Equation (A.17) was called by Fermi the 'golden rule' of quantum mechanics. It is usually applied to a system like that shown in Fig. A.4,

Fig. A.4. Energy levels for the base-states of a system which progresses from state a to an ensemble of states b.

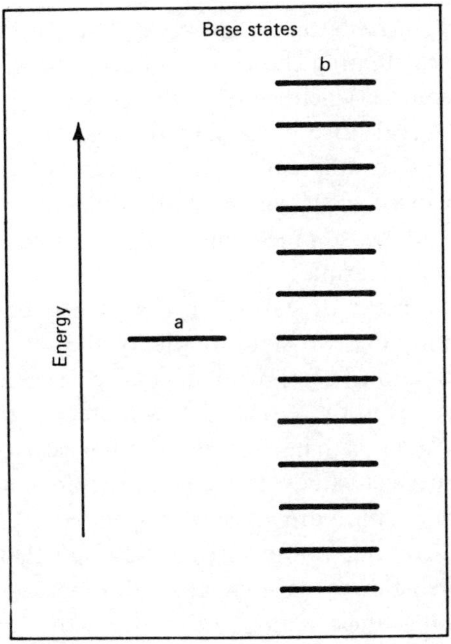

† Pointed out to me by Professor Kenneth Stolarsky.

and derived by perturbation theory. In Fig. A.4 we see a multi-state system with one state labelled a and an array of states labelled b, covering a spread of energies. The matrix elements to go from a to any of the bs are assumed to all have the same value, $-A$. The energy difference between successive b states is ρ^{-1}. Thus ρ is the density of b states to which a can change.

The perturbation calculation starts with the system known to be in state a and confines itself to times so small that a is not appreciably depleted. Thus each b state interacts with a practically full a state and no account need be taken of what the other b states are doing. Thus the occupation of the b states proceeds, for the short time, as if we had independent 2-state systems like Fig. A.3. An equation like equation (A.12) for $|C_b|^2$ applies to each b state, but the base-states shown in Fig. A.4. are used instead of the eigenstates of the two-state systems. This means that $r^2 + 1$ is replaced by r^2 and ω_{21} is defined as $(H_{bb} - H_{aa})/\hbar$ instead of $(E_2 - E_1)/\hbar$. Instead of equation (A.15) one gets

$$\frac{\mathrm{d}}{\mathrm{d}t}(\Sigma C_b{}^2) = \frac{4A^2\rho}{\hbar}\int_0^\infty \frac{\sin(Et/\hbar)}{E}\,\mathrm{d}E. \tag{A.18}$$

This integral is found in standard tables and leads also to equation (A.17).

(*a*) *Extension to exponential time-course.* However, both of the above approaches are wrong for calculating the system of Fig. A.4 at any but very small t, for two reasons: (1) neither uses the correct eigenvalues for the system of Fig. A.4, and (2) both square the amplitudes before summing the different paths (to different b states in the array). If it is impossible in principle to distinguish which of the different paths the transition takes, they should be summed (and thus allowed to show interference effects) before the squaring.

A correct treatment of an array of states like Fig. A.4, good for any length of time, would abandon the two-state simplicity and set up simultaneous equations like equations (A.1) with as many equations and unknown Cs as there are states in the system. For a finite, not too large, number of states the equations can be solved by numerical methods, finding and using the true eigenvalues and eigenstates for the whole system. Details of such a calculation are given in Appendix, Section A.7.

Fig. A.5 shows a gradual computer approach to the result that the full treatment not only gives the golden rule slope at small times but follows an exponential course at long times. Thus, to the extent that an actual system approximates the model of Fig. A.4, one need have no qualms about using the golden rule formula for the first-order rate constant which can be applied to long time intervals, even though the usual derivation by perturbation theory does not support it.

Other results of the study may be summarized as follows. The parameter $R = N/(A\rho)^2$, where N is the number of b states assumed in the array of Fig. A.4, $A = |H_{ab}|$ and ρ is the density of b states per unit energy interval, is useful. R is the square of the ratio of the energy interval N/ρ, covered by the spread of b states, to the energy splitting, $AN^{\frac{1}{2}}$, that would be produced by N b-states all having the same energy as the a state, since the total amplitude of their interaction with a would be $N^{\frac{1}{2}}$ times the interaction, A, of each with state a. If $R < 1$ the spread of the b states is not larger than the splitting which they produce between two dominating eigenstates by their interaction with a. The result is that the b states act together like a single b state and produce oscillatory behaviour approaching that of Fig. A.2 A. For $R > 1$ the oscillatory behaviour diminishes and

Fig. A.5. Time-course of tunnelling to an array of states calculated using eigenvalues and eigenvectors. The systems calculated are represented by Fig. A.4. The parameters used are: Curve 1: N = the number of b states in the array $= 21$, $|H_{ab}|\rho = 2$, $R = N/(|H_{ab}|\rho)^2 \approx 5$; curve 2: $N = 21$, $|H_{ab}|\rho = 1$, $R = 21$; curve 3: $N = 99$, $|H_{ab}|\rho = 1$, $R = 99$; curve 4: $N = 99$, $|H_{ab}|\rho = 0.5$, $R \approx 400$.

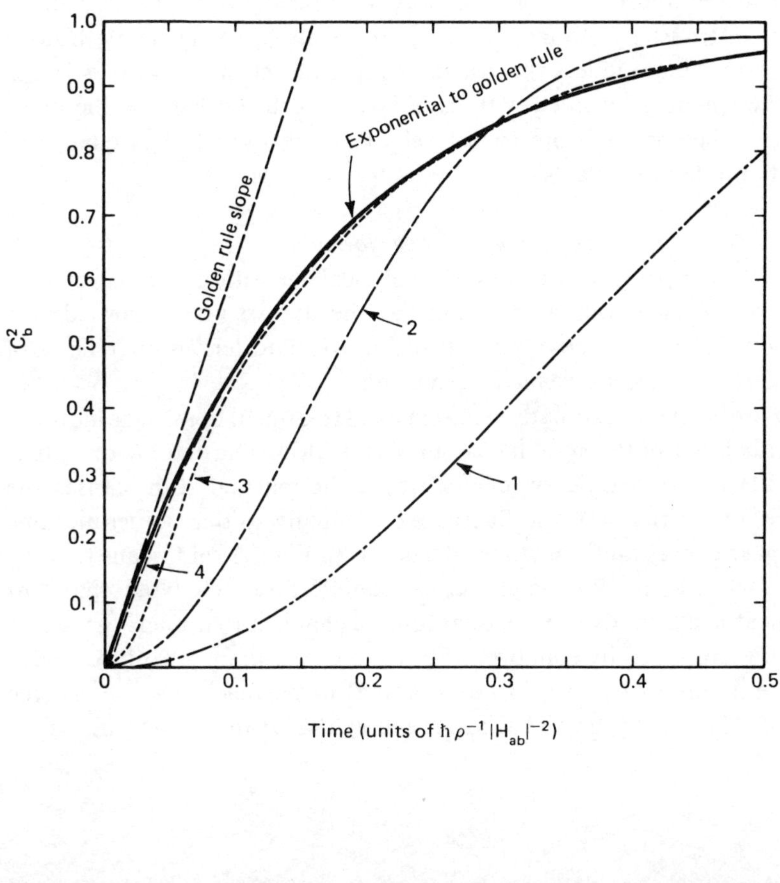

the bulk of a goes over into the array b and stays there. In order to approach the golden rule curve within a few per cent it is necessary that R equal several hundred. For the curves in Fig. A.5 the R values are given in the figure legend.

Another rule derived from the study is that to achieve stable occupation of the b array for long periods of time it is necessary that $A\rho$, the number of b states in the energy interval, A, must be large compared to 1. Without this, the initial following of the golden-rule curve might be quite good (if R is large) but after some initial period large departures suddenly appear.

In real systems the density of states, ρ, will not be as uniform as pictured in Fig. A.4. An array of states can arise from the many vibrational, rotational, etc., levels accompanying each electronically distinct state. Each of these levels is spread out further over a statistical distribution of energies because of thermal fluctuations and because of the uncertainty principle due to finite lifetimes, and possibly other fluctuations.

If we had a single level broadened only by the uncertainty principle we would find that the shorter the lifetime of the b state (due to decay to some state, c) the less would be the density, ρ, because of the greater spread in the uncertainty of its energy. The result would be slowing of the rate of transfer from a to b. The chances of observing such a bizarre effect seem remote unless the spacing between levels in an array is larger than the spread associated with each level, because otherwise the overlaping of the spreads from each level will mask the effect of change of spread of individual states.

A.1.4 *Derivation of the Landau–Zener formula*

Zener (1932) gives an exact analytical derivation of equation (5.8) using Weber functions as solutions to the approximate Schroedinger equation. The following is a simplified, approximate derivation in terms of the two-state formulas we have been using.

The problem is to estimate, using classical treatment of nuclear motions, the probability of the system changing from electronic state R to state P (Fig. 3.1) in a dynamic system in which thermal agitation causes the energies of the two states to fluctuate continually so that sometimes one has higher energy and sometimes the other. In Fig. 3.1 this means that the nuclear coordinate fluctuates across configuration C. Note that this treatment is alternative to the quantum-mechanical treatment of nuclear motion exemplified by equations (5.38) and (5.58). In our two-box model, R corresponds to particle in box a and P to particle in box b. To the right of C in Fig. 3.1, box b has lower energy than a and to the left, higher.

The change of state, being radiationless, can only occur when the energies are the same. For a given rate of passage from b having higher energy to b having lower energy than a, what is the probability the particle will transfer from a to b in a single pass?

Imagine the array of a, b boxes of Fig. A.3. We start out with the particle in a in each. Each box-pair has a gate which initially closes the path from a to b. Starting with the box-pair having the highest b-box, its gate is opened for an interval Δt allowing the possibility of the particle in its a to tunnel the barrier to b. After Δt the gate closes and the gate in the next box-pair down opens for an interval Δt. Successively each box-pair experiences open gate for a period Δt. This simulates the case of a single box-pair in which the energy of the b-box, H_{bb}, varies past H_{aa} with the speed:

$$\mathrm{d}(H_{bb} - H_{aa})/\mathrm{d}t = D/\Delta t, \tag{A.19}$$

where D is the energy spacing between successive box-pairs, if the total probability of transfer from a to b in the course of one passage is much less than unity, so that the depletion of occupation of a is negligible and it does not matter whether different a-boxes are used or whether a single a-box is exposed to a varying b-box. Since we are talking about total probabilities much less than one (the 'non-adiabatic' case) the only way to observe them would be to repeat the run many times and count the fraction of runs in which any particle transfers from a to b.

The net result will be the same as if all the gates were opened simultaneously for Δt and then closed. Thus the total probability, P_{ab}, will be the rate given by equation (A.17) multiplied by Δt:

$$P_{ab} = W_{ab}\Delta t = (2\pi/\hbar)|H_{ab}|^2 \Delta t/D, \tag{A.20}$$

where ρ is replaced by $1/D$. If the nuclear coordinate in Fig. 3.1 A is X, the difference in slope between the R and P curves in the neighbourhood of X_C is

$$\mathrm{d}(H_{aa} - H_{bb})/\mathrm{d}X = \Delta F. \tag{A.21}$$

Equations (A.19) to (A.21) with rate of change of nuclear coordinate = $\mathrm{d}X/\mathrm{d}t = v$ give

$$P_{ab} = (2\pi/\hbar)|H_{ab}|^2/v\Delta F, \tag{A.22}$$

which is the Landau–Zener formula, equation (5.8) with R, P changed to a, b, in the form suitable for the non-adiabatic case, that is, $P_{ab} \ll 1$.

Zener's more accurate result, good for the adiabatic case also is:

$$P_{ab} = 1 - \exp\{-(2\pi/\hbar)|H_{ab}|^2/v\Delta F\}. \tag{A.23}$$

Equation (A.22) is an approximation to (A.23) when $P_{ab} \ll 1$. As in the case of the golden rule, the exact solution forms an exponential extension to the approximate solution.

If P_{12} is the probability of crossing from state 1 to state 2 in a single passage through X_C in Fig. 3.1 B then it can be seen that:

$$P_{12} = 1 - P_{ab}. \tag{A.24}$$

The limitations and some extensions of the Landau–Zener formula have been discussed by Nikitin (1968), who also gives the formula

$$P = 2P_{12}(1 - P_{12}) = 2(1 - P_{ab})P_{ab}, \tag{A.25}$$

as the probability for change of state to have occurred upon double passage (round trip) through the region X_C. Nikitin also reviews a wide variety of cases such as those in which the R and P lines of Fig. 3.1 have different slants and curvatures and cases in which the forces on the atom cause the system starting out from one side of Fig. 3.1 to turn around at various places and go back.

Desouter–Lecomte & Lorquet (1977, 1979) have derived equation (A.24) with (A.23) and similar formulas for other cases by a general method based on the non-adiabaticity operator (see next section).

A.2 The Born–Oppenheimer approximation

To analyse a quantum-mechanical system one sets up a set of base-states for the system. The easiest to work with is the set of stationary states or eigenstates, that is, those whose Hamiltonian matrix is diagonal. It is practically impossible to do this exactly for a complicated molecular system with many nuclei and many electrons. A great simplification results, however, if one can treat the electronic motion separately from the nuclear motions. This is the objective of the Born–Oppenheimer approximation (Born & Oppenheimer, 1927; Born & Huang, 1954). It depends upon the fact that the electrons are much lighter and faster moving than the nuclei and so can follow the nuclei around and adjust to a good approximation to whatever configuration the nuclei have at a given moment almost as if the nuclei were standing still at that configuration. To a first approximation the method gives a reasonable separation of electronic and nuclear wave-functions. To a second approximation it shows that the separation is not complete but gives another method of approaching the problem of estimating the matrix elements for change from one state to another. I found the discussions by Kubo (1952), Kubo & Toyozawa (1955), Hobey & McLachlan (1960) and Bixon & Jortner (1968) useful. The following explanation mostly follows that of Hobey & McLachlan with some notation taken from Bixon & Jortner and some ideas from all of these.

The Hamiltonian of the system is taken as:

$$H = T_q(q) + T_Q(Q) + U(q, Q) + V(Q), \tag{A.26}$$

where q and Q represent the sets of electronic and nuclear coordinates, respectively; U and V are potential-energy functions of the coordinates indicated and T_q and T_Q are kinetic-energy operators as follows:

$$\left.\begin{aligned} T_q &= -\sum_j (\hbar^2/2m)(\partial^2/\partial q_j^{\,2}), \\ T_Q &= -\sum_k (\hbar^2/2M_k)(\partial^2/\partial Q_k^{\,2}). \end{aligned}\right\} \tag{A.27}$$

Here j and k enumerate the individual electronic and nuclear coordinates respectively, m is the electronic mass and M_k the reduced masses corresponding to the respective nuclear coordinates. $M_k^{\frac{1}{2}}$ is often incorporated into the definition of Q_k. Wave-functions are sought in the form

$$\psi_{nj}(q, Q) = \phi_n(q, Q)\chi_{nj}(Q). \tag{A.28}$$

The first step is to find the electronic eigenfunctions for the system with fixed nuclear coordinates. The part of H which depends on q is used:

$$(T_q + U)\phi_n(q, Q) = E_n(Q)\phi_n(q, Q). \tag{A.29}$$

Although Q is held constant it is clear that the eigenvalues E_n and the eigenfunctions ϕ_n which are solutions of this Schroedinger equation are dependent upon the particular value of Q as indicated in the parentheses. The ϕ_n are orthogonal and normalized for integration over dq at any given value of Q:

$$\int \mathrm{d}q\,\phi_m{}^*\phi_n = \delta_{mn}. \tag{A.30}$$

The second step is to find the nuclear wave-functions corresponding to a particular electronic state. Assuming that the electrons follow the nuclei adiabatically, that is that they continue to obey equation (A.29) without changing the value of n as Q varies, the electronic energy $E_n(Q)$ is used as a potential energy for the nuclear motion. This simple theory would give as the Schroedinger equation for the nuclear motion:

$$[T_Q + V + E_n]\chi_{nj}(Q) = W_{nj}{}'\chi_{nj}(Q). \tag{A.31}$$

However, the nuclear wave-function χ_{nj} and energy W_{nj}' for vibrational state j given by this equation are not very satisfactory, because if one applies the complete Hamiltonian to them, the average energy does not come out W_{nj}' and there are non-vanishing matrix elements between different vibrational states i and j belonging to the same electronic state n. Thus, matrix elements of the complete Hamiltonian between vibrational states of the same electronic state would be:

$$H_{ninj} = \iint \mathrm{d}q\, \mathrm{d}Q \psi_{ni}{}^* H \psi_{nj} = \iint \mathrm{d}q\, \mathrm{d}Q \phi_n{}^* \chi_{ni}{}^* \times (T_q + U + V + T_Q)\phi_n \chi_{nj}. \quad \text{(A.32)}$$

By equation (A.29), $(T_q + U)$ operating on ϕ_n can be replaced by E_n. $T_q(q)$ does not operate on $\chi_{nj}(Q)$. The operators U and V are multiplicative so that the order of their operations on ϕ and χ does not matter. T_Q operating on $\phi_n \chi_{nj}$ becomes

$$T_Q(Q)\phi_n(q, Q)\chi_{nj}(Q) = -\sum_k \frac{\hbar^2}{2M_k}\left[\phi_n \frac{\partial^2 \chi_{nj}}{\partial Q_k{}^2} + 2\frac{\partial \phi_n}{\partial Q_k}\frac{\partial \chi_{nj}}{\partial Q_k} + \frac{\partial^2 \phi_n}{\partial Q_k{}^2}\chi_{nj}\right]$$
$$= \phi_n T_Q \chi_{nj} + \chi_{nj} T_Q \phi_n - 2\sum_k \frac{\hbar^2}{2M_k}\frac{\partial \phi_n}{\partial Q_k}\frac{\partial \chi_{nj}}{\partial Q_k}. \quad \text{(A.33)}$$

Next we substitute these into equation (A.32) and perform the integration over dq. Assuming that ϕ_n is always real, the integral

$$\int \mathrm{d}q \phi_n{}^* \frac{\partial \phi_n}{\partial Q_k} = 0 \quad \text{(A.34)}$$

because then $\phi^* \partial\phi/\partial Q = \phi \partial\phi^*/\partial Q$ (and $\partial(\int \mathrm{d}q \phi^* \phi)/\partial Q = 0$). The rest of the terms become

$$H_{ninj} = \int \mathrm{d}Q \chi_{ni}{}^* \left\{ \int \mathrm{d}q \phi_n{}^* \phi_n [E_n + V + T_Q] + \int \mathrm{d}q \phi_n{}^* T_Q \phi_n \right\} \chi_{nj}$$
$$= \int \mathrm{d}Q \chi_{ni}{}^* [E_n + V + T_Q] \chi_{nj} + \int \mathrm{d}Q \chi_{ni}{}^* \langle n | T_Q | n \rangle \chi_{nj} \quad \text{(A.35)}$$
$$= W_{nj}{}' \delta_{ij} + \int \mathrm{d}Q \chi_{ni}{}^* \langle n | T_Q | n \rangle \chi_{nj}, \quad \text{(A.36)}$$

where $\langle n | T_Q | n \rangle$ is used as a symbol for $\int \mathrm{d}q \phi_n{}^* T_Q \phi_n$. The last term in (A.36) constitutes a matrix element between vibrational states i and j if $i \neq j$ and a deviation from $W_{nj}{}'$ if $i = j$.

The problem is remedied if equation (A.31) for the nuclear wavefunction is replaced by:

$$[T_Q + V + E_n + \langle n | T_Q | n \rangle] \chi_{nj}(Q) = W_{nj} \chi_{nj}(Q), \quad \text{(A.37)}$$

(called the 'diagonal correction'). Starting with equation (A.35) one can see that the matrix elements between vibrational states in the same electronic state now become

$$H_{ninj} = \int \mathrm{d}Q \chi_{ni}{}^* [T_Q + E_n + V + \langle n | T_Q | n \rangle] \chi_{nj} = W_{nj} \delta_{ij}. \quad \text{(A.38)}$$

We still do not have true eigenfunctions for the whole system. There are non-vanishing matrix elements between different electronic states. The Born–Oppenheimer approximation is to neglect them. These non-Born–Oppenheimer matrix elements may be analysed as follows. For $m \neq n$:

$$\begin{aligned} H_{\text{min}j} &= \iint \mathrm{d}q\,\mathrm{d}Q\phi_m{}^*\chi_{mi}{}^*[T_q + U + V + T_Q]\phi_n\chi_{nj} \\ &= \int \mathrm{d}Q\chi_{mi}{}^* \int \mathrm{d}q\phi_m{}^*[E_n + V + T_Q]\phi_n\chi_{nj} \\ &= \int \mathrm{d}Q\chi_{mi}{}^*\left\{\int \mathrm{d}q\phi_m{}^*\phi_n[E_n + V + T_Q] + \int \mathrm{d}q\phi_m{}^*T_Q\phi_n \right. \\ &\quad \left. - 2\sum_k \frac{\hbar^2}{2M_k}\int \mathrm{d}q\phi_m{}^*\frac{\partial\phi_n}{\partial Q_k}\frac{\partial}{\partial Q_k}\right\}\chi_{nj} \\ &= \int \mathrm{d}Q\chi_{mi}{}^*\left\{\langle m|T_Q|n\rangle - 2\sum_k \frac{\hbar^2}{2M_k}\left\langle m\left|\frac{\partial}{\partial Q_k}\right|n\right\rangle\frac{\partial}{\partial Q_k}\right\}\chi_{nj}, \end{aligned} \tag{A.39}$$

where equations (A.29), (A.30), and (A.33) have been used; $\langle m|T_Q|n\rangle$ stands for $\int \mathrm{d}q\phi_m{}^*T_Q\phi_n$; and $\langle m|\partial/\partial Q_k|n\rangle$ stands for

$$\int \mathrm{d}q\phi_m{}^*(\partial/\partial Q_k)\phi_n.$$

Equation (A.39) can also be written:

$$H_{\text{min}j} = \iint \mathrm{d}Q\,\mathrm{d}q\psi_{mi}{}^*\mathscr{L}\psi_{nj}, \tag{A.40}$$

where

$$\mathscr{L}\psi_{nj} = -\sum_k \frac{\hbar^2}{M_k}\left[\frac{\chi_{nj}}{2}\frac{\partial^2\phi_n}{\partial Q_k{}^2} + \frac{\partial\phi_n}{\partial Q_k}\frac{\partial\chi_{nj}}{\partial Q_k}\right] \tag{A.41}$$

and $\mathscr{L}$ is called the 'non-adiabaticity operator'. In all the applications discussed in Section 5.5.2 the term containing the second derivative is considered negligible. The factor $(\partial\phi_n/\partial Q_k)$, when averaged over the initial and final electronic wave-functions ϕ_a and ϕ_b, is designated $\langle a|\partial/\partial Q_k|b\rangle$ and, summed over the normal modes, it comprises the electronic part of the non-adiabaticity matrix element. If considered sufficiently constant against changing nuclear coordinates, it is factored out, as in the Condon approximation, which factors out H_{ab} from the total matrix element (see Section 5.3.2 and equations (5.21), (5.38) and

(5.41)). The factor $(\partial\chi_{nj}/\partial Q_k)$ generates derivatives of the Franck–Condon overlap integrals.

The electronic parts of H_{minj}, namely $\langle m|T_Q|n\rangle$ and $\langle m|\partial/\partial Q_k|n\rangle$ are small when $\phi(q, Q)$ varies only slowly with Q and the electronic energies E_m and E_n are not close together. Then spontaneous transitions from one electronic state to another are slow and the set of base-states, ψ_{nj}, are quite good approximations to eigenstates. Exact wave-functions, Ψ, for particular cases can be formed by linear combinations and will change with time:

$$\Psi(t) = \sum_{n,j} a_{nj}(t)\psi_{nj}. \tag{A.42}$$

It can be shown by expansion of $\partial\phi_n/\partial Q$ in terms of ϕ_m and by other manipulations of equation (A.29) that the matrix element $\langle m|\partial/\partial Q_k|n\rangle$ can be expressed as follows (Hobey & McLachlan, 1960):

$$\left\langle m\left|\frac{\partial}{\partial Q_k}\right|n\right\rangle = \frac{\langle m|\partial U(q, Q)/\partial Q|n\rangle}{E_n(Q) - E_m(Q)}, \tag{A.43}$$

where $\langle m|\partial U/\partial Q|n\rangle$ means $\int \mathrm{d}q\phi_m{}^*(\partial U/\partial Q)\phi_n$. This shows how it can become large if E_n approaches E_m closely as at the crossover points in Fig. 3.1. The other m to n coupling term, $\langle m|T_Q|n\rangle$, can be written in terms of $\langle m|\partial/\partial Q|n\rangle$:

$$\langle m|T_Q|n\rangle = -\sum_k \frac{\hbar^2}{2M_k}\left[\sum_l \left\langle m\left|\frac{\partial}{\partial Q_k}\right|l\right\rangle\left\langle l\left|\frac{\partial}{\partial Q_k}\right|n\right\rangle + \frac{\partial}{\partial Q_k}\left\langle m\left|\frac{\partial}{\partial Q_k}\right|n\right\rangle\right]. \tag{A.44}$$

In setting up the Born–Oppenheimer electronic base-states by solution of equation (A.29) at fixed nuclear coordinates we normally expect to arrive at the states 1 and 2 pictured in Fig. 3.1 B or Fig. A.1 G. As described in Section A.1 these are split apart by exchange interactions between R and P (or a and b) expressed by H_{RP} or H_{ab}. The matrix element H_{minj}, equation (A.39) or (A.40), then refers to jumping between curves 1 and 2. Adiabatic electron transfer is a transition from R to P over curve 1 without any quantum jump. It is most probable when H_{RP} is strong, the splitting between 1 and 2 is large, and H_{minj} is weak. The rate is governed entirely by the nuclear motion. Note that in the 'inverted' case ($\Delta E > \lambda$) (see Fig. 5.4, bottom) curve 1 does not have a double minimum and adiabatic transfer from R to P has no meaning.

Although a jump between adiabatic states (as 1 and 2 of Fig. 3.1 B) is not ordinarily an electron transfer such jumps are of interest because they may represent important electronic transitions. An example would be photo-isomerism of retinal in rhodopsin. The important nuclear coordinate is then the rotation about the $C_{11} = C_{12}$ bond. The ground state has a double minimum, one minimum corresponding to the *cis*-configuration and the other to the *trans*. An excited electronic state has a minimum energy at a bond rotation about halfway between *cis* and *trans* so that the diagram of energy v. bond rotation looks very much like the curves 1 and 2 of Fig. 3.1 B. Photoexcitation of the molecule to curve 2 is followed by a quantum jump, probably enabled by the non-adiabaticity operator, to the ground state which relaxes to either the *cis* or the *trans* configuration. For references, see Rosenfeld *et al.* (1977), Govindjee, Becher & Ebrey (1978) and Warshel (1979).

Electron transfer, however, in biological systems seems more often to be non-adiabatic. In this case the electron usually does not transfer when the nuclear coordinates go through the configuration C in Fig. 3.1. The exchange interactions are slower than the nuclear motion and during the time of passage through C there is not enough time to establish a splitting into 1 and 2. As already discussed in Section 5.5.2, it is then more appropriate in setting up the Born–Oppenheimer states to neglect the exchange (or 'configuration') interaction and to use diabatic R and P curves, Fig. 3.1 A, as the base-states. Then $H_{\mathrm{min}j}$ would refer to jumping from R to P. Although it is not always clear this is probably implied in papers which use the non-adiabaticity operator in non-adiabatic electron transfer contrary to the impression given in my earlier paper (DeVault, 1980). As suggested in Section 5.5.2, the exchange interaction that was neglected in setting up diabatic Born–Oppenheimer states should then be considered along with the neglected non-adiabaticity operators in forming the matrix elements for transition between the states. I think there is a need for theorists to address the question of the relative magnitude of the exchange interactions and the non-adiabaticity operators in furthering non-adiabatic electron transfer. The work of Desouter-Lecomte & Lorquet (1977, 1979) is an approach to the subject.

I think it will be worth while to present some basic results which illuminate the relation between adiabatic and diabatic formulations. First a little on terminology: Azumi & Matsuzaki (1977) review the subject and point out that 'Born–Oppenheimer' is used in several, sometimes contradictory ways. They suggest more careful distinction between three types of adiabatic ('AD') approximations: (1) the Born–Huang ('BH') which includes the 'diagonal correction' (equations (A.29) and (A.37)), (2) the Born–Oppenheimer ('BO') which does not (equations (A.29) and (A.31)), and (3) the 'crude Born–Oppenheimer' ('CBO') or crude adiabatic

('CA') which is obtained from equation (A.29) except that Q is held at some reference configuration, Q_0, and not solved for any other nuclear configuration. The resulting wave-functions do not depend on Q. However, they are useful as a basis set for expansion into either adiabatic or diabatic wave-functions in which the expansion coefficients depend upon Q. The next paragraph describes this in more detail.

The electronic Hamiltonian used in equation (A.29) when solved for CA eigenfunctions, ϕ_n°, and eigenvalues, E_n° is $[T_q(q) + U(q, Q_0)]$. This differs from the exact electronic Hamiltonian, $[T_q(q) + U(q, Q)]$, by $\Delta U(q, Q) = U(q, Q) - U(Q, Q_0)$, which then becomes the perturbation operator for the expansion:

$$\phi_n(q, Q) = \phi_n^\circ(q, Q_0) + \sum_{m \neq n} c_{mn}(Q)\phi_m^\circ(q, Q_0). \tag{A.45}$$

The coefficients are, by first-order perturbation theory:

$$c_{mn}(Q) = \frac{\langle m|\Delta U(q, Q)|n\rangle}{E_n^\circ(Q_0) - E_m^\circ(Q_0)}, \tag{A.46}$$

where $\langle m|\Delta U(q, Q)|n\rangle$ stands for $\int \phi_m^{\circ *} \Delta U \phi_n^\circ \mathrm{d}q$. This is often called the Herzberg–Teller expansion (Herzberg & Teller, 1933).

Azumi & Matsuzaki also discuss the term 'vibronic coupling'. They point out that it is sometimes, rather inappropriately, applied to the Herzberg–Teller expansion. The products of electronic and vibrational wave-functions (equation (A.28)) obtained by any of the three approximations, CA, BO or BH, represent what can be called 'vibronic states'. Vibronic coupling is then the coupling between these states expressed by equation (A.42) (which can also be written without the dependence on time for a given instant of time) to give exact, non-adiabatic states of the system.

Another way of forming exact states, not used much in practice, but illuminating theoretically, is to expand them in terms of the electronic wave-functions with coefficients which are functions of Q:

$$\Psi_k = \sum_n \phi_n(q, Q)\xi_{nk}(Q) \tag{A.47}$$

or

$$\Psi_k = \sum_n \phi_n^\circ(q, Q_0)\xi_{nk}^\circ(Q). \tag{A.48}$$

The coefficients, ξ, can be found by solving the following coupled differential equations which are obtained by applying the full Schroedinger equation to Ψ_k, multiplying on the left by ϕ_m^* or $\phi_m^{\circ *}$ and integrating over q:

$$\begin{aligned}[] [T_Q(Q) + V(Q) + E_m(Q) + \langle m|T_Q|m\rangle - E_k]\xi_{mk}(Q) \\ + \sum_{n \neq m}\left[\langle m|T_Q|n\rangle - \sum_j \frac{\hbar^2}{M_j}\left\langle m\left|\frac{\partial}{\partial Q_j}\right|n\right\rangle\frac{\partial}{\partial Q_j}\right]\xi_{nk}(Q) = 0 \end{aligned} \tag{A.49}$$

or:

$$[T_Q(Q) + V(Q) + E_m{}^{\circ}(Q_0) + \langle m|\Delta U|m\rangle - E_k]\xi_{mk}{}^{\circ}(Q) \\ + \sum_{n \neq m} \langle m|\Delta U|n\rangle \xi_{nk}{}^{\circ}(Q) = 0. \qquad \text{(A.50)}$$

The different equations in each coupled set are obtained by varying m over all possible values. If the coupling terms in these equations, those involving $n \neq m$, are neglected we have the Schroedinger equations in the adiabatic approximations for the vibrational wave-functions ($\xi(Q) \rightarrow \chi(Q)$) and vibronic energy values ($E_k \rightarrow W_{mk}$); equation (A.49) becomes (A.37) and (A.50) becomes that appropriate to CA wave-functions. If the coupling terms in (A.49) with the ξ_{nk} replaced by the χ_{nk} are multiplied on the left by $\chi_{mi}{}^*$ and integrated over dQ we obtain equation (A.39) for the matrix elements for transition or coupling between the adiabatic states.

Siebrand and co-workers (Gregory *et al.*, 1976; Laplante & Siebrand, 1978) have studied simple, two-state models with a single nuclear co-ordinate which can be solved exactly in order to establish the relationship between adiabatic and diabatic approaches to vibronic coupling in the same model. They developed formulas which hold for all degrees of coupling from weak to strong. These will be an extension of the standard two-state system introduced in Sections A.1.1 and A.1.2.

They represent the diabatic states directly by CA wave-functions without any expansion, since there is no need for them to depend upon Q in the model under consideration. According to equation (A.50), with the coupling terms neglected, the potential-energy function for nuclear motion is $[V(Q) + E_m{}^{\circ}(Q_0) + U_{mn}(Q)]$, where U_{mn} will be used as an abbreviation for $\langle m|\Delta U|n\rangle$. Whether the CA wave-functions lead to diabatic or adiabatic states depends upon whether or not these functions cross at some value of Q. Since we will choose crossing energy surfaces, I will call the states 'DB' rather than 'CA' in order to emphasize their diabatic nature.

The relation between the AD and the DB wave-functions is given by equations (A.11) and may also be written in matrix form:

$$\begin{pmatrix} \phi_1 \\ \phi_2 \end{pmatrix} = \begin{pmatrix} \cos\theta & \sin\theta \\ -\sin\theta & \cos\theta \end{pmatrix} \begin{pmatrix} \phi_b{}^{\circ} \\ \phi_a{}^{\circ} \end{pmatrix}, \qquad \text{(A.51)}$$

where $\sin\theta = A_1 = -B_2$ and $\cos\theta = A_2 = B_1$. This transformation is 'unitary', meaning (since its elements are real) that its transpose is its reciprocal. From equations (A.11) we also have:

$$\tan 2\theta = \frac{2}{\cot\theta - \tan\theta} = \frac{1}{r} = \frac{2H_{ab}}{H_{bb} - H_{aa}}. \qquad \text{(A.52)}$$

θ is zero at $Q = Q_0$ and $\pi/4$ at the crossing point.

The matrix elements for the electronic Hamiltonian are calculated from the wave-functions by the formula:

$$H_{mn} = \int \phi_m{}^*[T_q(q) + U(q, Q)]\phi_n \, \mathrm{d}q. \tag{A.53}$$

For the AD basis set, using equation (A.29) this gives equation (A.6) with $H_{11} = E_1(Q)$ and $H_{22} = E_2(Q)$. For the DB basis set it is more convenient to write the formula:

$$H_{mn} = \int \phi_m{}^{\circ *}[T_q(q) + U(q, Q_0) + \Delta U(q, Q)]\phi_n{}^\circ \, \mathrm{d}q. \tag{A.54}$$

This gives equation (A.2), where, with the DB equivalent of equation (A.29):

$$\left.\begin{aligned} H_{aa} &= E_a{}^\circ + U_{aa}(Q), \\ H_{bb} &= E_b{}^\circ + U_{bb}(Q), \\ H_{ab} &= H_{ba} = U_{ab}(Q). \end{aligned}\right\} \tag{A.55}$$

So far the equations apply to any shape of potential-energy function, except that only three of the four functions can be chosen arbitrarily.

We next fix our model by choosing specific potential-energy functions. For simplicity we will make the diabatic curves harmonic, each with the same frequency and choose $\Delta E = 0$. Thus:

$$\left.\begin{aligned} H_{aa}(Q) &= E_a{}^\circ + V(Q) + U_{aa}(Q) = \tfrac{1}{2}M\omega^2(Q + \Delta Q/2)^2, \\ H_{bb}(Q) &= E_b{}^\circ + V(Q) + U_{bb}(Q) = \tfrac{1}{2}M\omega^2(Q - \Delta Q/2)^2. \end{aligned}\right\} \tag{A.56}$$

Since in this chapter we are using Q to represent the nuclear coordinate, we now use ΔQ to represent the displacement of vibrational equilibrium position represented earlier by Q.

The AD surfaces are then fixed by the choice, $H_{ab} = U_{ab} = -A$, and are given by equations (A.9) which may also be written:

$$\left.\begin{aligned} H_{11} &= E_1(Q) = \tfrac{1}{2}M\omega^2[(\Delta Q)^2/4 + Q^2] - [M^2\omega^4Q^2(\Delta Q)^2/4 + A^2]^{\frac{1}{2}} \\ H_{22} &= E_2(Q) = \tfrac{1}{2}M\omega^2[(\Delta Q)^2/4 + Q^2] - [M^2\omega^4Q^2(\Delta Q)^2/4 + A^2]^{\frac{1}{2}} \end{aligned}\right\} \tag{A.57}$$

From equations (A.56) and (A.52) we find the relation between Q and r or θ:

$$r = \frac{M\omega^2\Delta Q}{2A} \cdot Q = \cot 2\theta. \tag{A.58}$$

The solution of the Schroedinger equation for nuclear motion in the DB case (equation (A.50) with U_{mn} neglected) with the potentials of equation (A.56) yields harmonic oscillator wave-functions $\chi_{av}{}^\circ$ and $\chi_{bw}{}^\circ$ with energies $(v + \frac{1}{2})\hbar\omega$ and $(w + \frac{1}{2})\hbar\omega$. $\chi_{av}{}^\circ$ will be centred at $Q = -\Delta Q/2$ and $\chi_{bw}{}^\circ$ at $Q = +\Delta Q/2$.

In the AD case, the potentials (A.57) are not harmonic and the solution of the equation (A.37) for nuclear motion is more complicated, particularly in the neighbourhood of $Q = 0$, corresponding to the crossover point for the diabatic curves. At larger absolute values of Q the AD curves approach the DB curves and so the nuclear motion should be nearly harmonic near the bottoms of the minima.

On the other hand, if we go to the exact equations (A.47) and (A.48) and write them for our two-state model we have:

$$\phi_1 \xi_{1k} + \phi_2 \xi_{2k} = \phi_a{}^\circ \xi_{ak}{}^\circ + \phi_b{}^\circ \xi_{bk}{}^\circ \tag{A.59}$$

Substituting for ϕ_1 and ϕ_2 by equation (A.51) we get:

$$\begin{aligned}\phi_a{}^\circ \xi_{ak}{}^\circ + \phi_b{}^\circ \xi_{bk} &= (\phi_b{}^\circ \cos\theta + \phi_a{}^\circ \sin\theta)\xi_{1k} + (-\phi_b{}^\circ \sin\theta + \phi_a{}^\circ \cos\theta)\xi_{2k} \\ &= \phi_a{}^\circ(\xi_{2k}\cos\theta + \xi_{1k}\sin\theta) + \phi_b{}^\circ(-\xi_{2k}\sin\theta + \xi_{1k}\cos\theta),\end{aligned}$$

$$\left.\begin{aligned} \text{or} \quad & \begin{pmatrix}\xi_{ak}{}^\circ \\ \xi_{bk}{}^\circ\end{pmatrix} = \begin{pmatrix}\cos\theta & \sin\theta \\ -\sin\theta & \cos\theta\end{pmatrix}\begin{pmatrix}\xi_{2k} \\ \xi_{1k}\end{pmatrix}, \\ \text{or} \quad & \begin{pmatrix}\xi_{2k} \\ \xi_{1k}\end{pmatrix} = \begin{pmatrix}\cos\theta & -\sin\theta \\ \sin\theta & \cos\theta\end{pmatrix}\begin{pmatrix}\xi_{ak}{}^\circ \\ \xi_{bk}{}^\circ\end{pmatrix}. \end{aligned}\right\} \tag{A.60}$$

These express exact relations between the ξs. The ξs are functions of Q but they are not exact eigenfunctions of the nuclear Schroedinger equation. However, if the term U_{mn} in equation (A.50) ($= U_{ab}$ or A, in our model) is very small, the $\xi_{mk}{}^\circ$ become very close to the harmonic oscillator wave-functions $\chi_{mk}{}^\circ$, where m is a or b and k is v or w. Then equation (A.60) expresses the AD ξ in terms of combinations of pairs of these harmonic oscillator wave-functions:

$$\begin{aligned}\xi_{2k} &\approx \chi_{ak}{}^\circ \cos\theta - \chi_{bk}{}^\circ \sin\theta, \\ \xi_{1k} &\approx \chi_{ak}{}^\circ \sin\theta + \chi_{bk}{}^\circ \cos\theta.\end{aligned} \tag{A.61}$$

The matrix elements for the complete Hamiltonian are calculated on the DB basis as follows.

For diagonal elements:

$$\begin{aligned}H_{mvmw} &= \iint \mathrm{d}q\, \mathrm{d}Q \phi_m{}^{\circ *} \chi_{mv}{}^{\circ *} H \phi_m{}^\circ \chi_{mw}{}^\circ \\ &= \int \mathrm{d}Q \chi_{mv}{}^{\circ *}[T_Q + E_m{}^\circ(Q_0) + U_{mm}]\chi_{mw}{}^\circ \\ &= (v + \tfrac{1}{2})\hbar\omega\delta_{vw},\end{aligned} \tag{A.62}$$

where $m = a$ or b.

For the off-diagonal elements:

$$H_{avbw} = \int \mathrm{d}Q \chi_{av}{}^{\circ *} U_{ab} \chi_{bw}{}^{\circ} = -AC(v, w), \tag{A.63}$$

where $C(v, w)$ is the overlap integral given by equation (5.36), or, in the notation used here:

$$C(v, w) = \int \chi_{av}{}^{\circ *} \chi_{bw}{}^{\circ} \, \mathrm{d}Q \tag{A.64}$$

and $-A$ is the assumed constant value of the electronic matrix element, H_{ab}. Note the relationship between equations (A.63) and (5.38).

In the AD basis the diagonal elements are given by equation (A.38) and the off-diagonal by (A.39). By substituting DB electronic wavefunctions for the AD functions ϕ_1 and ϕ_2 through the relation (A.51) we find:

$$\begin{aligned}\left\langle 1 \left| \frac{\partial}{\partial Q} \right| 2 \right\rangle &= \int (\phi_b{}^{\circ *} \cos\theta + \phi_a{}^{\circ *} \sin\theta) \\ &\quad \times \left(-\phi_b{}^{\circ} \frac{\mathrm{d} \sin\theta}{\mathrm{d}Q} + \phi_a{}^{\circ} \frac{\mathrm{d}\cos\theta}{\mathrm{d}Q} \right) \mathrm{d}q \\ &= -\cos\theta \frac{\mathrm{d}\sin\theta}{\mathrm{d}Q} + \sin\theta \frac{\mathrm{d}\cos\theta}{\mathrm{d}Q} = -\frac{\mathrm{d}\theta}{\partial Q} \\ &= \frac{\alpha}{2(1+\alpha^2 Q^2)},\end{aligned} \tag{A.65}$$

where $\alpha = M\omega^2 \Delta Q / 2A$ and equation (A.58) was used.
Similarly:

$$\begin{aligned}\langle 1 | T_Q | 2 \rangle &= -\frac{\hbar^2}{2M} \int \phi_1{}^* \frac{\partial^2}{\partial Q^2} \phi_2 \, \mathrm{d}q = \frac{\hbar^2}{2M} \frac{\partial^2 \theta}{\partial Q^2} \\ &= \frac{\hbar^2}{2M} \frac{\alpha^3 Q}{(1+\alpha^2 Q^2)^2}\end{aligned} \tag{A.66}$$

and

$$\langle 1 | T_Q | 1 \rangle = \langle 2 | T_Q | 2 \rangle = \frac{\hbar^2}{8M} \cdot \frac{\alpha^2}{(1+\alpha^2 Q^2)^2}. \tag{A.67}$$

Note that the functions described by equations (A.65)–(A.67) approach zero at values of Q larger than $1/\alpha$. This means that the important values of Q in evaluating $H_{\mathrm{min}j}$ by equation (A.39) are those close to $Q = 0$, especially if A is small. Desouter–Lecomte & Lorquet (1977, 1979) make use of this fact to develop a simple method of evaluating H_{minj} approximately. They avoid finding the wave-functions χ_{mi}^* and χ_{nj} (which are not harmonic in the adiabatic case) by treating the nuclear motion classically as was done by Landau and Zener (see the preceding section). They advocate use of the non-adiabaticity operator because molecular orbital calculations yield the adiabatic potential surfaces first. The diabatic surfaces are not so well defined and must be obtained by further calculation.

A.3 Charge transfer or intervalence spectra

Mulliken (1952) provided the quantum-mechanical basis for the strong colours often noted when a colourless electron acceptor is put with a colourless electron donor. The absorption band accompanying the formation of a complex between the two is called a 'charge-transfer' band because it arises directly from the possibility of electron movement between acceptor and donor. The term 'intervalence' band is also used (Hush, 1967; Allen & Hush, 1967).

The quantum-mechanical states involved are well illustrated by our discussion of the two-box problem. Suppose the state *b* ('electron in box *b*') has lower energy than state *a* ('electron in *a*'). Box *b* is the acceptor and *a* the donor, Fig. A.1 C. This corresponds to some abscissa to the right of the centre of Fig. A.1 G. The amount of *a* and *b* included in eigenstates 1 and 2 are given by equations (A.11) and illustrated in Fig. A.1 F. The energies are given by equations (A.9). These are the same as Mulliken's equations if overlap of *a* with *b* is neglected.

Overlap comes into Mulliken's equations because he represents his states with Schroedinger wave-functions, which we did not use in discussing the two-box problem, but did use in the H_2^+ discussion. However, in the non-adiabatic cases in which we are interested the overlap integrals are indeed negligible.

In any case the states 1 and 2 constructed from *a* and *b* are eigenfunctions of the coupled system and therefore orthogonal to each other. In wave-function form they are represented in Fig. A.6. ψ_2 corresponds to what would be the antisymmetrical state if the energies of *a* and *b* were equal, and, therefore has a node in the barrier region. ψ_1 does not.

The electronic jump from state 1 to state 2 is indicated by CT in Fig. 3.1, where '*R*' and '*P*' are used for '*a*' and '*b*', respectively. The strength of absorption of electromagnetic radiation is proportional to a matrix element, μ_{12}, representing the radiation dipole. This is an oscillating electric dipole produced by electric charge oscillating between two positions in a manner analogous to the oscillation of electric current in a radio or television antenna. In the case in which we are interested the oscillation is along the axis between the centres of a and b and it is effective on radiation that is polarized with electric vector in this direction. The quantum-mechanical calculation is illustrated in Fig. A.6. The axial direction is taken as x. The lowest diagram represents ψ_1. It is shown as having most amplitude in the region of b, on the right, since we are taking the energy of b as much lower than that of a. See also Fig. A.1 F. The diagram above it represents $x \cdot \psi_1$. At the centre, x is equal to zero, and is negative to the left. The next diagram up represents ψ_2. It has most amplitude in the vicinity of a, in accordance also with Fig. A.1 F. The top diagram represents $\psi_2 x \psi_1$, the product of the two diagrams just below it. The dipole matrix element in the x direction is the area of the top diagram multiplied by the electronic charge. Thus:

$$\mu_{X_{12}} = e \int \psi_1 x \psi_2 \, \mathrm{d}\tau. \tag{A.68}$$

The fact that it comes out negative as drawn in the diagram is irrelevant. The absolute value is important and the fact that it is not zero. It could have come out zero if ψ_2 had the same symmetry character as ψ_1.

The value of the dipole matrix element can be seen from Fig. A.6 to be approximately an amount of charge at a equal to $A_1 A_2$ and an equal and opposite charge at b (equal to $B_1 B_2$), in units of electronic charge, multiplied by the distance between a and b. Thus, using equations

Fig. A.6. Steps in calculation of charge-transfer dipole matrix element, $\mu_{X_{12}}$. The values of A_1, A_2, B_1 and B_2 are given by equations (A.11).

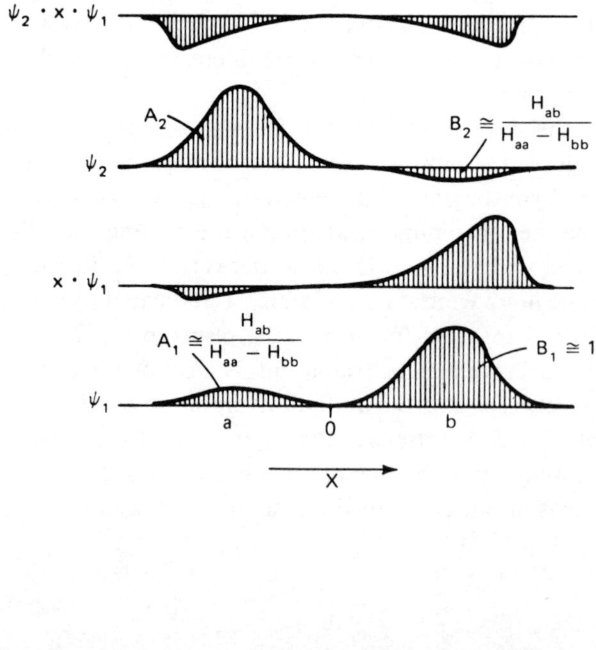

(A.11):

$$\mu_{X_{12}} \approx e \frac{|H_{ab}|}{H_{aa} - H_{bb}} r_{ab}, \tag{A.69}$$

where e is the electronic charge and r_{ab} is the distance between the centres of a and b. This is Mulliken's (1952) equation (20) with his equation (11) and neglect of overlap.

The 'oscillator strength', f, defined as $2m\omega\mu_{X_{12}}{}^2/\hbar e^2$ (see Merzbacher, 1961, p. 446; Herzberg, 1950, p. 383) is then:

$$f = \frac{2m\omega|H_{ab}|^2 r_{ab}{}^2}{\hbar(H_{aa} - H_{bb})^2}. \tag{A.70}$$

(Note: some authors, e.g. Herzberg, insert the factor 3 into the denominator of f, but then the dipole moment used corresponds to the sum of the squares of all three directional components). f is dimensionless. The Einstein A-coefficient for spontaneous emission would be:

$$A_{ab} = \frac{2e^2\omega^2}{mc^3} \cdot f = \frac{4e^2\omega^3|H_{ab}|^2 r_{ab}{}^2}{\hbar c^3(H_{aa} - H_{bb})^2} \tag{A.71}$$

in Gaussian units or

$$A_{ab} = \frac{e^2\omega^2}{2\pi mc^3\varepsilon_0} \cdot f = \frac{e^2\omega^3|H_{ab}|^2 r_{ab}{}^2}{\pi\hbar c^3\varepsilon_0(H_{aa} - H_{bb})^2} \tag{A.72}$$

in SI units. A_{ab} is a first-order rate constant with units: s^{-1}.

Experimentally, f is measured by integrating the absorption spectrum over all bands corresponding to the same electronic transition. In terms of the molecular absorption *cross-section*, $\sigma(\omega)((\text{cm}^2 \text{ or } \text{m}^2) \text{ molecule}^{-1})$:

$$f = \underset{\text{(e.s.u.)}}{\frac{mc}{2\pi^2 e^2}\int \sigma(\omega)\,d\omega} = \underset{\text{(SI units)}}{\frac{2mc\varepsilon_0}{\pi e^2}\int \sigma(\omega)\,d\omega}. \tag{A.73}$$

In terms of the molar *extinction coefficients*, $\varepsilon(\omega)$ $(\text{l mol}^{-1}(\text{cm}^{-1} \text{ or } \text{m}^{-1}))$:

$$f = \underset{\text{(e.s.u.)}}{\frac{mc \times 10^3 \ln 10}{2\pi^2 e^2 N_0}\int \varepsilon(\omega)\,d\omega} = \underset{\text{(SI units)}}{\frac{2mc\varepsilon_0 \times 10^3 \ln 10}{\pi e^2 N_0}\int \varepsilon(\omega)\,d\omega}, \tag{A.74}$$

where N_0 = Avogadro's number, and ln 10 = 2.3026.

A.4 Interpretation of Franck–Condon factors

It may be interesting to note that the Franck–Condon factors resulting from overlaps of harmonic oscillator wave-functions found, for example, in equations (5.37), (5.61) and (5.66)–(5.68), can be interpreted in terms of Poisson statistics. The general formula for Poisson statistics is:

$$P_n = \frac{R^n e^{-R}}{n!}, \tag{A.75}$$

where P_n is the probability of n hits on a target in a 'trial', and R is the average rate of hits per trial. A random process is assumed. The $\exp(-R)$ may be considered a normalization factor and any expression proportional to $R^n/n!$ will fit the form.

Suppose an electron-transfer process has energy loss ΔE, such that $\Delta E/\hbar\omega = p$ vibrational quanta must be absorbed by the surroundings in order to balance the energy. Suppose that there are N oscillators in the surroundings capable of taking or giving the vibrational quanta and that N is a large number, very much larger than p. We may expect, then, that any given oscillator will not take or give more than one quantum. We need to count the ways in which the net number of quanta absorbed by the surroundings is p. Any distribution in which q oscillators lose one quantum each and $q + p$ gain one quantum each will do.

The average rate at which a given oscillator will absorb a quantum is proportional to $C^2(n, n + 1)$ for that oscillator, where n is the number of vibrational quanta it already has. See equation (5.36). $C^2(n, n + 1)$ is different for each oscillator because they have different values for n and because $C^2(n, n + 1)$, according to equation (5.37) is a function of S, and S is different for each since they are at different distances from the location of the electron transfer and have different couplings. However, before we finish we will form suitable averages of both n and S. With some loss of rigor in the argument let us assume an average value for $C^2(n, n + 1)$ and apply it to all oscillators, now.

From equation (A.75) the probability of having $(q + p)$ oscillators gain one quantum each is:

$$P_{q+p} = C^2(n, n + 1)^{q+p} \exp(-C^2(n, n + 1))/(q + p)! \tag{A.76}$$

In a similar way the probability that q oscillators will lose one quantum each is

$$P_q = C^2(n, n - 1)^q \exp(-C^2(n, n - 1))/q! \tag{A.77}$$

The total rate of electron transfer should then be proportional to the joint probability, $P_{q+p}P_q$, of both occurring summed over all possible values of q:

$$W = A \sum_q C^2(n, n + 1)^{q+p} C^2(n, n - 1)^q \exp(-C^2(n, n + 1) - C^2(n, n - 1))/(q + p)!q! \tag{A.78}$$

With some modifications we are following the arguments used by

Huang & Rhys (1950). They were interested in values of S much smaller than 1, and derived formulas for oscillator wave-functions with equilibrium position displaced, by a Taylor's series expansion of the undisplaced wave-function in powers of the displacement. They used these to evaluate $C^2(n, n')$ by equation (5.36), retaining only the lowest order terms. We can get the same results by taking the limit to which equation (5.37) goes as S becomes smaller. The result is:

$$\left.\begin{aligned} C^2(n, n+1) &\approx (n+1)S, \\ C^2(n, n-1) &\approx nS. \end{aligned}\right\} S \ll 1. \tag{A.79}$$

Note that these results conform to the general observation (Feynman *et al.*, 1965) that Bose particles (in this case vibrational quanta) tend to go into or leave a situation with a rate proportional to the number of such particles already there, or that will be there, whichever is larger.

If one substitutes equations (A.79) into (A.78) one obtains with a little rearrangement:

$$W = A \exp(-S(2n+1))\left(\frac{n+1}{n}\right)^{\frac{1}{2}p} \sum_q \frac{S^{2q+p}(n+1)^{q+\frac{1}{2}p} n^{q+\frac{1}{2}p}}{q!(q+p)!}. \tag{A.80}$$

It can be seen that the summation is the modified Bessel function, equation (5.62), namely, $I_p(2S[n(n+1)]^{\frac{1}{2}})$. With proper evaluation of A and interpretation of n as $\bar{n}$ we have equation (5.67). Although this derivation assumes that S is small, the previous derivation shows that the same formula holds for S large.

The formula (5.37) for $C^2(n, n')$ can also be interpreted in terms of Poisson distributions. It may be rewritten as:

$$C^2(n, n+p) = \mathrm{e}^{-S}\left[\sum_{i=0}^{n} (-1)^i \left\{\left(\frac{n!}{i!(n-i)!}\right)\frac{S^i}{i!}\right\}^{\frac{1}{2}} \times \left\{\left(\frac{(n+p)!}{(p+i)!(n-i)!}\right)\frac{S^{p+i}}{(p+i)!}\right\}^{\frac{1}{2}}\right]^2. \tag{A.81}$$

An interpretation could be:

Consider the number of ways by which an oscillator may change from having n vibrational quanta to having $n+p$ quanta. It could simply gain p quanta, or it could lose 1 and gain $p+1$, or lose 2 and gain $p+2$, etc. These possibilities are indicated by the summation over i. The quantity under the first radical is the number of ways (combinations) of losing i quanta out of n quanta times a factor, $S^i/i!$, proportional to the Poisson probability of having i losses, given that the average rate is S. (For this calculation we do not seem to have the Bose–Einstein n factor operating and we slur over the normalization factor.) We put this probability under a radical because at this point we want probability *amplitudes* (quantum-mechanical). The $(-1)^i$ factor indicates that the amplitudes for the losses should have negative phase factor. The second radical has under it the number of combinations of $p+i$ quanta taken from the total of $n+p$ quanta that the oscillator ends up with, multiplied by the Poisson probability of $p+i$ gains at average rate S. All these amplitudes are summed before squaring because they represent different courses which are operationally indistinguishable.

A.5 The electron barrier in a polar medium

The following is a derivation of equation (5.107) used in constructing Fig. 5.7. Our system consists of two electron sites, Do and Ac (for donor and acceptor), at a centre to centre distance D from each other in a medium, M. The radii of the electron sites and of the molecules of the medium will all be assumed, for simplicity, to be equal to a single value, r. For further simplicity we assume that the oxidized Do and Ac sites and, of course, the molecules of the medium are electrically neutral. If the oxidized redox sites are charged, this will simply add a constant pattern of polarization to everything that follows. To get the curves of Fig. 5.7, we must calculate the energy of an electron introduced from outside into either Do or Ac or into M at various positions along a line between the two centres.

The electrical potential that would be observed in a vacuum at a distance d from a charge q is

$$\phi_v = \frac{q}{4\pi\varepsilon_0 d}, \tag{A.82}$$

where ε_0 is the permittivity of free space to give SI units. In a medium with optical dielectric constant κ_0, immediately after placing the charge and before the nuclei had a chance to move in response to the charge the potential would be:

$$\phi_0 = \frac{q}{4\pi\varepsilon_0 d\kappa_0}. \tag{A.83}$$

κ_0 is equal to the square of the refractive index of the medium. After the nuclei have moved to new equilibrium positions the potential would be:

$$\phi = \frac{q}{4\pi\varepsilon_0 d\kappa}, \tag{A.84}$$

where κ is the total (low-frequency) dielectric constant. This potential can be thought of as made up of a potential ϕ_v due to the charge at distance

d in a vacuum, a potential ϕ_e due to the electronic polarization induced in the medium by the charge and a potential ϕ_N due to the nuclear polarization induced in the medium by the charge:

$$\phi = \phi_v + \phi_e + \phi_N. \tag{A.85}$$

The potential due to electronic polarization is:

$$\phi_e = \phi_0 - \phi_v = \frac{q}{4\pi\varepsilon_0 d}\left(\frac{1}{\kappa_0} - 1\right) = \frac{-q}{4\pi\varepsilon_0 d\kappa_e}. \tag{A.86}$$

That due to the nuclear polarization is:

$$\phi_N = \phi - \phi_0 = \frac{q}{4\pi\varepsilon_0 d}\left(\frac{1}{\kappa} - \frac{1}{\kappa_0}\right) = \frac{-q}{4\pi\varepsilon_0 d\kappa_N}. \tag{A.87}$$

These equations define:

$$\frac{1}{\kappa_e} = 1 - \frac{1}{\kappa_0}, \tag{A.88}$$

$$\frac{1}{\kappa_N} = \frac{1}{\kappa_0} - \frac{1}{\kappa}. \tag{A.89}$$

To get the energy of an electron introduced into the system in a particular fixed nuclear polarization one needs the vertical electron affinity, the energy released when the electron falls into the system from infinity without any change in the nuclear positions. Thus for nuclear configuration A in Figs. 3.1 and 5.1, or curve A in Fig. 5.7, the electron energy will be in the donor:

$$E_{Do}(A) = -V_{Do} - \frac{e^2}{4\pi\varepsilon_0 r\kappa_e} - \frac{e^2}{4\pi\varepsilon_0 r\kappa_N}, \tag{A.90}$$

in the medium:

$$E_M(A) = -V_M - \frac{e^2}{4\pi\varepsilon_0 r\kappa_e} - \frac{e^2}{4\pi\varepsilon_0 x\kappa_N}, \tag{A.91}$$

and in the acceptor:

$$E_{Ac}(A) = -V_{Ac} - \frac{e^2}{4\pi\varepsilon_0 r\kappa_e} - \frac{e^2}{4\pi\varepsilon_0 D\kappa_N}, \tag{A.92}$$

where V_i is the vertical electron affinity of each substance in a vacuum; the second term on the right-hand side of each equation gives the energy lowering due to the fast electronic polarization and is given by ϕ_e (equation A.86) × the electronic charge; the last term of each is the energy lowering due to the fixed nuclear configuration A and is given by ϕ_N (equation A.87) × the electronic charge: and x is the distance from the centre of Do along the line toward Ac. The electronic charge is $-e$ and this replaces q in (A.86) and (A.87). The appropriate values for d in the electronic term is r in each case. Nuclear configuration A is in equilibrium with the electron in Do so the appropriate values for d in the nuclear terms are r in (A.90), x in (A.91) and D in (A.92).

The curve B in Fig. 5.7 corresponding to nuclear polarization B in Figs. 3.1 and 5.1 is similarly expressed by:

$$E_{\mathrm{Do}}(B) = -V_{\mathrm{Do}} - \frac{e^2}{4\pi\varepsilon_0 r\kappa_{\mathrm{e}}} - \frac{e^2}{4\pi\varepsilon_0 D\kappa_{\mathrm{N}}}, \tag{A.93}$$

$$E_{\mathrm{M}}(B) = -V_{\mathrm{M}} - \frac{e^2}{4\pi\varepsilon_0 r\kappa_{\mathrm{e}}} - \frac{e^2}{4\pi\varepsilon_0 (D-x)\kappa_{\mathrm{N}}}, \tag{A.94}$$

$$E_{\mathrm{Ac}}(B) = -V_{\mathrm{Ac}} - \frac{e^2}{4\pi\varepsilon_0 r\kappa_{\mathrm{e}}} - \frac{e^2}{4\pi\varepsilon_0 r\kappa_{\mathrm{N}}}. \tag{A.95}$$

This is in equilibrium with the electron in Ac.

Nuclear polarization C in Figs. 3.1 and 5.1, curve C in Fig. 5.7, is not an equilibrium configuration. It is any random, temporary polarization that happens to give:

$$E_{\mathrm{Do}}(C) = E_{\mathrm{Ac}}(C). \tag{A.96}$$

Thus many different curves might do. However, the polarization fluctuations most likely to influence this result will be those in the neighbourhood of Do and Ac and so for simplicity I have chosen to represent them as polarizations that would be in equilibrium with fictitious charges q_{Do} and q_{Ac} at Do and at Ac, respectively, if there were such charges there. Once these charges have served their purpose to generate the polarization C, they will magically disappear, leaving C as a non-equilibrium polarization.

The curve C, Fig. 5.7, will be described by:

$$E_{\mathrm{Do}}(C) = -V_{\mathrm{Do}} - \frac{e^2}{4\pi\varepsilon_0 r\kappa_{\mathrm{e}}} + \frac{eq_{\mathrm{Do}}}{4\pi\varepsilon_0 r\kappa_{\mathrm{N}}} + \frac{eq_{\mathrm{Ac}}}{4\pi\varepsilon_0 D\kappa_{\mathrm{N}}}, \tag{A.97}$$

$$E_{\mathrm{M}}(C) = -V_{\mathrm{M}} - \frac{e^2}{4\pi\varepsilon_0 r\kappa_{\mathrm{e}}} + \frac{eq_{\mathrm{Do}}}{4\pi\varepsilon_0 x\kappa_{\mathrm{N}}} + \frac{eq_{\mathrm{Ac}}}{4\pi\varepsilon_0 (D-x)\kappa_{\mathrm{N}}}, \tag{A.98}$$

$$E_{\mathrm{Ao}}(C) = -V_{\mathrm{Ao}} - \frac{e^2}{4\pi\varepsilon_0 r\kappa_{\mathrm{e}}} + \frac{eq_{\mathrm{Do}}}{4\pi\varepsilon_0 D\kappa_{\mathrm{N}}} + \frac{eq_{\mathrm{Ac}}}{4\pi\varepsilon_0 r\kappa_{\mathrm{N}}}. \tag{A.99}$$

Equation (A.96) means that:

$$\frac{e}{4\pi\varepsilon_0 \kappa_{\mathrm{N}}}\left(\frac{q_{\mathrm{Do}}}{r} + \frac{q_{\mathrm{Ac}}}{D} - \frac{q_{\mathrm{Do}}}{D} - \frac{q_{\mathrm{Ac}}}{r}\right) = V_{\mathrm{Do}} - V_{\mathrm{Ac}}. \tag{A.100}$$

The energy required to produce polarization C from A with the electron in Do is $E^{\ddagger}$, that is:

$$E^{\ddagger} = E_{\mathrm{Do}}(C) - E_{\mathrm{Do}}(A). \tag{A.101}$$

Substituting from equations (A.97) and (A.90) gives:

$$E^{\ddagger} = \frac{e}{4\pi\varepsilon_0\kappa_{\mathrm{N}}}\left(\frac{q_{\mathrm{Do}}}{r} + \frac{q_{\mathrm{Ac}}}{D} + \frac{e}{r}\right). \tag{A.102}$$

Marcus (1965) calculated $E^{\ddagger}$ as the energy necessary to stretch oscillators in the medium to give the required polarization and arrived at equation (3.3) or, in terms of $E^{\ddagger}$ and ΔE, equation (5.7). We may note that the reorganization energy, λ, is given by:

$$\lambda = E_{\mathrm{Ac}}(A) - E_{\mathrm{Ac}}(B) = E_{\mathrm{Do}}(B) - E_{\mathrm{Do}}(A). \tag{A.103}$$

Substituting (A.92) and (A.95) or (A.93) and (A.90) into (A.103) gives:

$$\lambda = \frac{e^2}{4\pi\varepsilon_0\kappa_{\mathrm{N}}}\left(\frac{1}{r} - \frac{1}{D}\right) = \frac{e^2}{4\pi\varepsilon_0 D\kappa_{\mathrm{N}}}\left(\frac{1-R}{R}\right), \tag{A.104}$$

where $R = r/D$. With equation (A.89) this is seen to agree with equation (3.5) if $r_1 = r_2 = r$ and $r_{12} = D$. (We are assuming the two redox sites to be some distance apart while equation (3.5) assumes them to be in contact with each other.)

Note also that the net loss of electronic energy is given by:

$$\Delta E = E_{\mathrm{Do}}(A) - E_{\mathrm{Ac}}(B). \tag{A.105}$$

Substituting (A.90) and (A.95) into (A.105) gives:

$$\Delta E = V_{\mathrm{Ac}} - V_{\mathrm{Do}}. \tag{A.106}$$

From equations (A.100), (A.104) and (A.106) we have:

$$\frac{q_{\mathrm{Ac}} - q_{\mathrm{Do}}}{e} = \frac{4\pi\varepsilon_0 D\kappa_{\mathrm{N}}}{e^2}\left(\frac{R}{1-R}\right)\Delta E = \frac{\Delta E}{\lambda}, \tag{A.107}$$

and from equation (A.102) and (A.104):

$$\frac{q_{\mathrm{Do}}}{e} + \frac{q_{\mathrm{Ac}}}{e}R = (1-R)\frac{E^{\ddagger}}{\lambda} - 1. \tag{A.108}$$

Equations (A.107) and (A.108) with (5.7) yield:

$$\frac{q_{\mathrm{Ac}}}{e} = \frac{\Delta E - \lambda}{\lambda(1+R)} + \frac{E^{\ddagger}}{\lambda}\left(\frac{1-R}{1+R}\right) = \frac{(\Delta E - \lambda)}{4\lambda(1+R)}\left[\frac{\Delta E}{\lambda}(1-R) + 3 + R\right] \tag{A.109}$$

and

$$\frac{q_{\mathrm{Do}}}{e} = \frac{q_{\mathrm{Ac}}}{e} - \frac{\Delta E}{\lambda}. \tag{A.110}$$

The values for q_{Do} and q_{Ac} given by (A.109) and (A.110) are substituted into equations (A.97) to (A.99) to get a full description of curve C of Fig. 5.7.

For use in equation (5.97) one needs the electron barrier height above the energy level of the electron in curve C, that is, $E_M(C) - E_{Do}(C)$. From equations (A.97) and (A.98) with (A.104), (A.109) and (A.110) we have:

$$V(x) = A + B\frac{D}{x} + C\frac{D}{D-x}, \tag{A.111}$$

where

$$A = V_{Do} - V_M - E^{\ddagger} + \frac{\lambda}{(1-R)}$$

$$= V_{Do} - V_M - \frac{(\Delta E)^2}{4\lambda} + \frac{\Delta E}{2} + \frac{\lambda}{4}\left(\frac{3+R}{1-R}\right),$$

$$B = -\frac{R}{1-R^2}[\lambda - E^{\ddagger}(1-R) + R\Delta E]$$

$$= -\frac{R}{1-R^2}\left[\frac{\lambda}{4}(3+R) + \frac{\Delta E}{2}(1+R) - \frac{(\Delta E)^2}{4\lambda}(1-R)\right],$$

$$C = -\frac{R}{1-R^2}[\lambda - E^{\ddagger}(1-R) - \Delta E]$$

$$= -\frac{R}{1-R^2}\left[\frac{\lambda}{4}(3+R) - \frac{\Delta E}{2}(1+R) - \frac{(\Delta E)^2}{4\lambda}(1-R)\right].$$

Equation (A.111) is seen to be the same as equation (5.107).

If one remembers that usual electron tunnelling distances, D, are not many molecular diameters wide, it is clear that the electron barrier will be 'lumpier' than shown by Fig. 5.7 or by equations (A.97) to (A.99) and (A.111). These would correspond to averages in a statistical-mechanical sense.

Fig. 5.7 was drawn for the normal, strongly coupled case ($\lambda > \Delta E$) shown in Figs. 3.1 and 5.1. In the case $\lambda = \Delta E$, equations (A.109) and (A.111) show that $q_{Ac} = 0$ and $q_{Do} = -e$. Thus curve C is then identical with curve A as expected for this case. In the 'abnormal', or weakly coupled case, ($\lambda < \Delta E$), q_{Ac} becomes positive in value. This means that curve C will then reach upward above the level of the unpolarized medium as it approaches the position of the acceptor. The energy gap, ΔE, is then so large that a negative polarization around the position of the acceptor is necesary to raise the energy level of an electron in it sufficiently to match the level in the donor.

A.6 Formulas for frequency change

Section 5.5.1 called attention to the work of the Kakitani's who showed that the changes of vibrational frequencies which could accompany electron transfer may have effects which are as important in the excitation of vibrational modes as are the displacements of the equilibrium positions. We can expect, therefore, that, in the future, interest in the effects of frequency change will grow. In order to assist further investigation, some basic formulas are presented here.

A change of frequency upon electron transfer means that the two parabolas in Figs. 3.1 and 3.2 will not have the same shape. The one with higher frequency will be narrower in appearance and the vibrational energy levels in it will be further apart. The reorganization energy, λ, will be different depending upon which curve is used in its definition. If defined as the energy necessary to displace the system an amount Q, while in the R electronic state from nuclear configuration, A, then $\lambda = MQ^2\omega^2$ (see equations (5.3) and (5.4)). The corresponding definition in the P electronic state would be $\lambda' = MQ^2\omega'^2/2$ where ω is the angular frequency in the R state and ω' that in the P state. In terms of a mean frequency, $\bar{\omega} = 2\omega\omega'/(\omega + \omega')$, we could define $\bar{\lambda}$ as $MQ^2\bar{\omega}^2/2$ and $\bar{S}$ as $MQ^2\bar{\omega}/2\hbar$.

The following formula is derived from those given by Mannebach (1951) and applies to harmonic oscillators shifted in both position and frequency without approximation. It thus includes equation (5.37) as a special case (for $\gamma = 0$).

$$C^2(n, n') = n!n'!(1-\gamma)^{n'-n}(1-\gamma^2)^{n+\frac{1}{2}}\,\mathrm{e}^{-\bar{S}}\bar{S}^{n+n'} \times \left\{ \sum_{m=0}^{A}\sum_{q=0}^{B}\sum_{r=0}^{C}(-1)^{q+r}\frac{\gamma^m(1-\gamma)^{2r-m}(1-\gamma^2)^{-r}\bar{S}^{-m-q}}{q!(n-q)!(n'-q)!} \times f(m-r, n'-2m-q+2r)f(r, n-q-2r)\right\}^2 \tag{A.112}$$

The symbols not already defined for equation (5.37) have the following meanings: $\gamma = (\omega' - \omega)/(\omega + \omega')$ is a measure of the frequency shift; $A = \mathrm{int}(n/2) + \mathrm{int}(n'/2)$, where $\mathrm{int}(x)$ = largest integer equal to or less than x; $B = \min\ (n,n')$ = the lesser of n or n'; $C = \min(m, \mathrm{int}(n'-q)/2)$. The functions, $f(x, y)$ are defined as:

$$f(x, y) = \sum_{\text{all } v_i}\left(\prod_{u=1}^{x}(2u - 1 + v_u)\right), \tag{A.113}$$

where the summation is over all products containing different combinations of the whole numbers, v_i, in which $0 \leqslant v_1 \leqslant v_2 \ldots \leqslant v_x \leqslant y$. By definition, $f(0, y) \equiv 1$. For a systematic exploration of the combinations one can let v_1 go from 0 to y while v_2 goes from v_1 to y, and, in general, v_u goes from v_{u-1} to y if v_0 is defined as 0.

Equation (A.112) is in the form of a polynomial in powers of γ and of $(1-\gamma)$. If the equation is expanded in powers of γ to the first order, only, one obtains:

$$C^2(n, n')\,\mathrm{e}^{-\bar{S}}\bar{S}^{n+n'}n!n'!\left\{\sum_{q=0}^{B}(-1)^q\frac{\bar{S}^{-q}}{q!(n-q)!(n'-q)!}\right.$$
$$\times\left[1+\left(\frac{n-n'}{2}\right)\gamma+\frac{\gamma}{\bar{S}}\sum_{r=0}^{D}(-1)^r f(1-r, n'-2-q+2r)\right.$$
$$\left.\left.\times\, f(r, n'-q-2r)\right]\right\}^2, \qquad \text{(A.114)}$$

where the limit B is the same as in (A.112) and $D = \min(1, \operatorname{int}(\tfrac{1}{2}(n'-q)))$.

Koide (1960) has also derived a formula for $C^2(n, n')$ with both frequency and position shift. His equation (34), translated into our symbols, is:

$$C^2(n, n') = \frac{\exp(-\alpha^2)}{n!n'!}(-\gamma)^{n'}(1-\gamma^2)^{\frac{1}{2}}\left|\mathscr{D}^n\exp(-\gamma\alpha^2)H_{n'}(\alpha\xi)\right|^2, \qquad \text{(A.115)}$$

where the operator $\mathscr{D}^n = \sum_{k=0}^{n}\binom{n}{k}(-\alpha)^{n-k}(\partial/\partial\alpha)$; $\binom{n}{k}$ is the binomial coefficient, $n!/k!(n-k)!$; $\alpha^2 = MQ^2\omega/2\hbar = S$ as defined by equations (5.3), (5.4) and under (5.20), $= \bar{S}/(1+\gamma)$; and $H_{n'}(x)$ is the Hermitian polynomial of order n'. This formula still requires considerable work to expand the derivatives. Koide's formula for the case of no frequency change, his equation (13) squared, if corrected for a missing radical over $n!$ and $m!$ is identical to equation (5.37).

Going next to the more general formulation, which starts with equation (5.41), we rewrite equation (5.43) so as to allow for change of frequency:

$$(\mathrm{FC}) = \frac{1-\exp(-\beta\hbar\omega)}{2\pi\hbar}\int \mathrm{d}t\int \mathrm{d}x\int \mathrm{d}y\sum_{n=0}^{\infty}\chi_n(x)\chi_n(y)$$
$$\times\exp[-n(\beta\hbar+it)\omega]\sum_{n'=0}^{\infty}\chi_{n'}(x')\chi_{n'}(y')\exp(n'i\omega't)$$
$$\times\exp\left[i\left(\frac{\omega'-\omega}{2}-\frac{\Delta E}{\hbar}\right)t\right]. \qquad \text{(A.116)}$$

Note that in substituting for $E_{n'}$ and E_n in equation (5.41) we must now also include the zero-point vibrational energies, $\hbar\omega''/2$ and $\hbar\omega/2$. The expression corresponding to (5.48) becomes:

$$\sum_{n=0}^{\infty} \chi_n(x)\chi_n(y)\exp[-n(\beta\hbar+it)\omega]$$
$$=\left\{\frac{M\omega/\pi\hbar}{1-\exp[-2(\beta\hbar+it)\omega]}\right\}^{\frac{1}{2}}\exp\left\{\left[-v^2\tanh\left(\frac{(\beta\hbar+it)\omega}{2}\right)\right.\right.$$
$$\left.\left.-\frac{1}{4}w^2\coth\left(\frac{(\beta\hbar+it)\omega}{2}\right)\right]\frac{M\omega}{\hbar}\right\}. \tag{A.117}$$

That corresponding to 5.50 is:

$$\sum_{n'=0}^{\infty} \chi_{n'}(x')\chi_{n'}(y')\exp(n'i\omega' t)=\left\{\frac{M\omega'/\pi\hbar}{1-\exp(2i\omega' t)}\right\}^{\frac{1}{2}}$$
$$\times\exp\left\{\left[-(v-Q)^2\tanh\left(\frac{-i\omega' t}{2}\right)-\frac{1}{4}w^2\coth\left(\frac{-i\omega' t}{2}\right)\right]\frac{M\omega'}{\hbar}\right\}. \tag{A.118}$$

In place of Equation (5.51) we have:

$$(\mathrm{FC})=\frac{[1-\exp(-\beta\hbar\omega)]M(\omega\omega')^{\frac{1}{2}}}{2\pi^2\hbar^2}$$
$$\times\int \mathrm{d}t\frac{\exp[i(((\omega'-\omega)/2)-i\Delta E/\hbar)t]}{\{1-\exp[-2(\beta\hbar+it)\omega]\}^{\frac{1}{2}}\{1-\exp(2i\omega' t)\}^{\frac{1}{2}}}$$
$$\times\int \mathrm{d}v\exp\left\{\left[-v^2\omega\tanh\left(\frac{(\beta\hbar+it)\omega}{2}\right)\right.\right.$$
$$\left.\left.-(v-Q)^2\omega'\tanh\left(\frac{-i\omega' t}{2}\right)\right]\frac{M}{\hbar}\right\}$$
$$\times\int \mathrm{d}w\exp\left\{\left[-w^2\omega\coth\left(\frac{(\beta\hbar+it)\omega}{2}\right)\right.\right.$$
$$\left.\left.-w^2\omega'\coth\left(\frac{-i\omega' t}{2}\right)\right]\frac{M}{4\hbar}\right\}. \tag{A.119}$$

Expression (5.52), the integral over dw, becomes:

$$\int \mathrm{d}w=\left\{\frac{M}{4\pi\hbar}\left[\omega\coth\left(\frac{(\beta\hbar+it)\omega}{2}\right)+\omega'\coth\left(\frac{-i\omega' t}{2}\right)\right]\right\}^{-\frac{1}{2}}. \tag{A.120}$$

The integral over $\mathrm{d}v$ becomes, instead of equation (5.53):

$$\int \mathrm{d}v = \left\{\frac{M}{\pi\hbar}\left[\omega \tanh\left(\frac{(\beta\hbar + it)\omega}{2}\right) + \omega' \tanh\left(\frac{-i\omega' t}{2}\right)\right]\right\}^{-\frac{1}{2}}$$

$$\times \exp\left\{\frac{MQ^2}{\hbar}\left[\frac{-\omega\omega' \tanh((\beta\hbar + it)\omega/2)\tanh(-i\omega' t/2)}{\omega \tanh((\beta\hbar + it)\omega/2) + \omega' \tanh(-i\omega' t/2)}\right]\right\}. \tag{A.121}$$

Instead of (5.54) we have:

$$\left\{\frac{M}{4\pi\hbar}\left[\omega \coth\left(\frac{(\beta\hbar + it)\omega}{2}\right) + \omega' \coth\left(\frac{-i\omega' t}{2}\right)\right]\right\}^{-\frac{1}{2}}$$

$$\times \left\{\frac{M}{\pi\hbar}\left[\omega \tanh\left(\frac{(\beta\hbar + it)\omega}{2}\right) + \omega' \tanh\left(\frac{-i\omega' t}{2}\right)\right]\right\}^{-\frac{1}{2}}$$

$$= \frac{\pi\hbar}{M}[1 - \exp(-2(\beta\hbar + it)\omega)]^{\frac{1}{2}}[1 - \exp(2i\omega' t)]^{\frac{1}{2}}$$

$$\times \exp\left(\frac{\beta\hbar\omega}{2}\right)\exp\left(\frac{i(\omega - \omega')t}{2}\right)$$

$$\times \{(\omega^2 + \omega'^2)\sinh((\beta\hbar + it)\omega)\sinh(-i\omega' t)$$

$$+ 2\omega\omega'[\cosh((\beta\hbar + it)\omega)\cosh(-i\omega' t) - 1]\}^{-\frac{1}{2}}. \tag{A.122}$$

Substituting this into (A.120) and (A.121) and then into (A.119), with some manipulation, gives the equivalent of equation (5.55):

$$(\mathrm{FC}) = \frac{(\omega\omega')^{\frac{1}{2}}\sinh(\beta\hbar\omega/2)}{\pi\hbar(\omega + \omega')}\int \mathrm{d}t\left\{\sinh^2\left(\frac{\beta\hbar\omega + i(\omega - \omega')t}{2}\right)\right.$$

$$\left. - \left(\frac{\omega - \omega'}{\omega + \omega'}\right)^2 \sinh^2\left(\frac{\beta\hbar\omega + i(\omega + \omega')t}{2}\right)\right\}^{-\frac{1}{2}}$$

$$\times \exp\left\{\frac{-2\left(\dfrac{MQ^2\omega\omega'}{\hbar(\omega + \omega')}\right)\sinh\left(\dfrac{(\beta\hbar + it)\omega}{2}\right)\sinh\left(\dfrac{-i\omega' t}{2}\right)}{\sinh\left(\dfrac{\beta\hbar\omega + i(\omega - \omega')t}{2}\right) + \left(\dfrac{\omega - \omega'}{\omega + \omega'}\right)\sinh\left(\dfrac{\beta\hbar\omega + i(\omega + \omega')t}{2}\right)}\right.$$

$$\left. - i\frac{\Delta E}{\hbar}t\right\}. \tag{A.123}$$

If put into the form of equation (5.57) it becomes:

$$(\mathrm{FC}) = \frac{(1 - \gamma^2)^{\frac{1}{2}}}{2\pi\hbar}\int_{-\infty}^{\infty} \mathrm{d}t\{[1 + \bar{n}(1 - \exp(i(\omega' - \omega)t))]^2$$

$$- \gamma^2[(\bar{n} + 1)\exp(i\omega' t) - \bar{n}\exp(-i\omega t)]^2\}^{-\frac{1}{2}}$$

$$\times \exp\left\{ -\bar{S}\frac{1+\bar{n}[1+\exp(i(\omega'-\omega)t)]-(\bar{n}+1)\exp(i\omega' t)-\bar{n}\exp(-i\omega t)}{1+\bar{n}[1-\exp(i(\omega'-\omega)t)]+\gamma[(\bar{n}+1)\exp(i\omega' t)-\bar{n}\exp(i\omega t)]}\right.$$

$$\left. +i\left(\frac{\omega'-\omega}{2}-\frac{\Delta E}{\hbar}\right)t\right\}. \tag{A.124}$$

Both $\bar{S}$ and γ are the same as in equation (A.112). The mean vibrational quantum number $\bar{n}$, before electron transfer, is substituted for $[\exp(\beta\hbar\omega)-1]^{-1}$.

So far, no approximations have been made. However, equation (A.124) is not easily integrated without approximations. For the present let us seek an approximation good to the first order in $\omega'-\omega$ or γ. The second term in the pre-exponential part is already second order because of γ^2. For the first term we have:

$$1+\bar{n}[1-\exp(i(\omega'-\omega)t)] \approx 1-i\bar{n}(\omega'-\omega)t \approx \exp(-i\bar{n}(\omega'-\omega)t). \tag{A.125}$$

We thus are able to move the pre-exponential part into the exponent.

The denominator in the coefficient of $\bar{S}$ in the exponent can be moved to the numerator with the approximation: $(1+\varepsilon)^{-1} \approx 1-\varepsilon$, where $\varepsilon \ll 1$, to become:

$$1-\bar{n}[1-\exp(i(\omega'-\omega)t)]-\gamma[(\bar{n}+1)\exp(i\omega' t)-\bar{n}\exp(-i\omega t)]. \tag{A.126}$$

The coefficient of $\bar{S}$ becomes:

$$\begin{aligned}
1-\bar{n}^2 &+ \exp(i\omega' t)[\bar{n}^2-1-\gamma(\bar{n}+1)^2] \\
&+ \exp(-i\omega t)[\bar{n}(\bar{n}-1)+\gamma\bar{n}(\bar{n}+1)] \\
&+ \exp(2i\omega' t)[\gamma(\bar{n}+1)^2] \\
&+ \exp(-2i\omega t)[-\gamma\bar{n}^2] \\
&+ \exp(i(2\omega'-\omega)t)[-\bar{n}(\bar{n}+1)-\gamma\bar{n}(\bar{n}+1)] \\
&+ \exp(i(\omega'-2\omega)t)[-\bar{n}^2+\gamma\bar{n}^2] \\
&+ \exp(i(\omega'-\omega)t[2\bar{n}] \\
&+ \exp(2i(\omega'-\omega)t)[\bar{n}^2]
\end{aligned} \tag{A.127}$$

It contains not only the two fundamental frequencies, ω and ω', but also difference frequencies and harmonics.

Next, substitute $\tilde{\omega}(1-\gamma)$ for ω and $\tilde{\omega}(1+\gamma)$ for ω', where $\tilde{\omega}=(\omega+\omega')/2$. Equation (A.127) then becomes, with the approximate expansion of exp $(\pm mi\gamma\tilde{\omega}t)$ to $(1 \pm mi\gamma\tilde{\omega}t)$:

$$\begin{aligned}\text{Coeff. of } \bar{S} = {} & 2\bar{n} + 1 + \exp(i\tilde{\omega}t)(1 + i\gamma\tilde{\omega}t)[\bar{n}^2 - 1 - \gamma(\bar{n}+1)^2] \\ & + \exp(-i\tilde{\omega}t)(1 + i\gamma\tilde{\omega}t)[\bar{n}(\bar{n}-1) + \gamma\bar{n}(\bar{n}+1)] \\ & + \exp(2i\tilde{\omega}t)(1 + 2i\gamma\tilde{\omega}t)\gamma(\bar{n}+1)^2 \\ & + \exp(-2i\tilde{\omega}t)(1 + 2i\gamma\tilde{\omega}t)(-\gamma\bar{n}^2) \\ & + \exp(i\tilde{\omega}t)(1 + 3i\gamma\tilde{\omega}t)[-\bar{n}(\bar{n}+1) - \gamma\bar{n}(\bar{n}+1)] \\ & + \exp(-i\tilde{\omega}t)(1 + 3i\gamma\tilde{\omega}t)[-\bar{n}^2 + \gamma\bar{n}^2] \\ & + 4i\gamma\tilde{\omega}t(\bar{n}^2 + \bar{n}). \end{aligned} \tag{A.128}$$

Collecting terms, dropping those higher than first order in γ, and then approximating $\exp(i\tilde{\omega}t)i\gamma\tilde{\omega}t$ with $\exp(i\tilde{\omega}t)(\exp(i\gamma\tilde{\omega}t) - 1)$ and $\exp(-i\tilde{\omega}t)i\gamma\tilde{\omega}t$ with $\exp(-i\tilde{\omega}t)(1 - \exp(-i\gamma\tilde{\omega}t))$ gives the following form equivalent to equation (5.57) but now suitable for small changes of frequency:

$$\begin{aligned}(\text{FC}) \approx \frac{1}{2\pi\hbar}\int \mathrm{d}t \exp\Big\{ & -\bar{S}[2\bar{n} + 1 \\ & + \exp(i\tilde{\omega}t)(2\bar{n}+1) - \gamma(2\bar{n}+1)(\bar{n}+1) \\ & + \exp(-i\tilde{\omega}t)(-2\bar{n}(\bar{n}+1) + \gamma(2\bar{n}+1)\bar{n}) \\ & + \exp(2i\tilde{\omega}t)\gamma(\bar{n}+1)^2 \\ & + \exp(-2i\tilde{\omega}t)(-\gamma\bar{n}^2) \\ & + \exp(i(1+\gamma)\tilde{\omega}t)(-(2\bar{n}+1)(\bar{n}+1)) \\ & + \exp(-i(1+\gamma)\tilde{\omega}t)((2\bar{n}+1)\bar{n}) \\ & + 4i\gamma\tilde{\omega}t(\bar{n}+1)\bar{n}] + i\left[\gamma(2\bar{n}+1)\tilde{\omega} - \frac{\Delta E}{\hbar}\right]t\Big\}. \end{aligned} \tag{A.129}$$

This can now be put into a form integrable over t as follows:

The factor,

$$\begin{aligned}&\exp\{-2\bar{S}\bar{n}(\bar{n}+1)\mathrm{e}^{i\tilde{\omega}t} + 2\bar{S}\bar{n}(\bar{n}+1)\mathrm{e}^{-i\tilde{\omega}t}\} \\ &\quad = \sum_{m=-\infty}^{\infty} (-1)^m \mathrm{e}^{mi\tilde{\omega}t} J_m(4\bar{S}\bar{n}(\bar{n}+1)), \end{aligned} \tag{A.130}$$

where $J_m(x) =$ is the ordinary Bessel function ('of the first kind') of order m.

The factor

$$\begin{aligned}&\exp\{\bar{S}(2\bar{n}+1)(\bar{n}+1)\mathrm{e}^{i(1+\gamma)\tilde{\omega}t} - \bar{S}(2\bar{n}+1)\bar{n}\,\mathrm{e}^{-i(1+\gamma)\tilde{\omega}t}\} \\ &\quad = \sum_{l=-\infty}^{\infty} \left(\frac{\bar{n}+1}{\bar{n}}\right)^{l/2} \mathrm{e}^{li(1+\gamma)\tilde{\omega}t} J_l(2\bar{S}(2\bar{n}+1)[\bar{n}(\bar{n}+1)]^{\frac{1}{2}}). \end{aligned} \tag{A.131}$$

The factor,

$$\begin{aligned}&\exp\{-\bar{S}\gamma[-(2\bar{n}+1)(\bar{n}+1)\mathrm{e}^{i\tilde{\omega}t} + (2\bar{n}+1)\bar{n}\,\mathrm{e}^{-i\tilde{\omega}t} \\ &\quad + (\bar{n}+1)^2\,\mathrm{e}^{2i\tilde{\omega}t} - \bar{n}^2\,\mathrm{e}^{-2i\tilde{\omega}t}]\} \approx 1 - \bar{S}\gamma[-(2\bar{n}+1)(\bar{n}+1)\mathrm{e}^{i\tilde{\omega}t}\end{aligned}$$

$$+ (2\bar{n}+1)\bar{n}\,\mathrm{e}^{-i\tilde{\omega}t} + (\bar{n}+1)^2\,\mathrm{e}^{2i\tilde{\omega}t} - \bar{n}^2\,\mathrm{e}^{-2i\tilde{\omega}t}]$$

$$= \sum_{r=-2}^{2} u(r)\,\mathrm{e}^{ri\tilde{\omega}t}, \tag{A.132}$$

where:

$$u(-2) = \gamma\bar{S}\bar{n}^2$$
$$u(-1) = -\gamma\bar{S}\bar{n}(2\bar{n}+1)$$
$$u(0) = 1$$
$$u(1) = \gamma\bar{S}(\bar{n}+1)(2\bar{n}+1)$$
$$u(2) = -\gamma\bar{S}(\bar{n}+1)^2.$$

Substituting (A.130), (A.131) and (A.132) into (A.129) gives:

$$(\mathrm{FC}) = \frac{1}{2\pi\hbar}\exp\{-\bar{S}(2\bar{n}+1)\}\left[\sum_{m=-\infty}^{\infty}(-1)^m J_m(4\bar{S}\bar{n}(\bar{n}+1))\right]$$
$$\times\left[\sum_{l=-\infty}^{\infty}\left(\frac{\bar{n}+1}{\bar{n}}\right)^{l/2} J_l(2\bar{S}(2\bar{n}+1)[\bar{n}(\bar{n}+1)]^{\frac{1}{2}})\right]$$
$$\times\left[\sum_{r=-2}^{2} u(r)\right]\times\int_{-\infty}^{\infty}\exp\Big\{i[m+l(1+\gamma)+r$$
$$-4\gamma\bar{S}\bar{n}(\bar{n}+1)+\gamma(2\bar{n}+1)]\tilde{\omega}t - i\frac{\Delta E}{\hbar}t\Big\}\,\mathrm{d}t. \tag{A.133}$$

(A.133) is written for only one mode of oscillation. Everything, except the $-i\Delta Et/\hbar$ is to be repeated as a factor for each mode in a multimode system.

In the form of equation (5.58), (A.129) with (5.21) can be written:

$$W = \frac{2\pi}{\hbar}|H_{RP}|^2\frac{1}{2\pi\hbar}\int_{-\infty}^{\infty} f(t)\exp(-i\Delta Et/\hbar)\,\mathrm{d}t, \tag{A.134}$$

where

$$f(t) = \exp\{-G + G_+(t) + G_-(t) + G_{2+}(t) + G_{2-}(t)$$
$$+ G_{1+\gamma}(t) + G_{-1-\gamma}(t) + G_0(t)\},$$

$$G_+(t) = -\sum_k \bar{S}_k[2\bar{n}_k(\bar{n}_k+1) - \gamma_k(2\bar{n}_k+1)(\bar{n}_k+1)]\exp(i\tilde{\omega}_k t),$$

$$G_-(t) = \sum_k \bar{S}_k[2\bar{n}_k(\bar{n}_k+1) - \gamma_k(2\bar{n}_k+1)\bar{n}_k]\exp(-i\tilde{\omega}_k t),$$

$$G_{2+}(t) = -\sum_k \bar{S}_k\gamma_k(\bar{n}_k+1)^2\exp(2i\tilde{\omega}_k t),$$

$$G_{2-}(t) = \sum_k \bar{S}_k\gamma_k\bar{n}_k{}^2\exp(-2i\tilde{\omega}_k t),$$

$$G_{1+\gamma}(t) = \sum_k \bar{S}_k(2\bar{n}_k+1)(\bar{n}_k+1)\exp(i(1+\gamma_k)\tilde{\omega}_k t),$$

$$G_{-1-\gamma}(t) = -\sum_k \bar{S}_k(2\bar{n}_k + 1)\bar{n}_k \exp(-i(1+\gamma_k)\tilde{\omega}_k t),$$

$$G_0(t) = \sum_k [-4\bar{S}_k(\bar{n}_k + 1)\bar{n}_k + (2\bar{n}_k + 1)]i\gamma_k\omega_k t,$$

$$G = G_+(0) + G_-(0) + G_{2+}(0) + G_{2-}(0) + G_{1+\gamma}(0) + G_{-1-\gamma}(0) = \sum_k \bar{S}_k(2\bar{n}_k + 1).$$

The summations over k allow for multiple modes of vibration. $k = 1, 2, \ldots, N$, where N is the total number of vibrational modes. An implicit assumption is that the electron transfer does not prevent the modes after transfer from being identified with corresponding modes before transfer. This result leads to the Kakitani (Kakitani & Kakitani, 1981) formula if the γ_k in the coefficients of S_k are set equal to zero. In this respect they followed Lin (1966) whose formulation is discussed later in this section.

The integral over dt in equation (A.133), modified for multimodes, for a given set of values for m_k, l_k and r_k is, in place of equation (5.63).

$$\int_{-\infty}^{\infty} \exp\left\{ i\sum_k [m_k + l_k(1+\gamma_k) + r_k - 4\gamma_k\bar{S}_k\bar{n}_k(\bar{n}_k + 1) + \gamma_k(2\bar{n}_k + 1)]\tilde{\omega}_k t - i\frac{\Delta E}{\hbar}t \right\} \mathrm{d}t$$

$$= 2\pi\delta\left(\sum_k [m_k + l_k(1+\gamma_k) + r_k - 4\gamma_k\bar{S}_k\bar{n}_k(\bar{n}_k + 1) + \gamma_k(2\bar{n}_k + 1)]\tilde{\omega}_k - \Delta E/\hbar \right). \tag{A.135}$$

In analogy to the procedure giving equations (5.64) and (5.65) we may isolate the vibrational mode of lowest frequency (still reasonably well coupled to the electron transfer), let the value for its k be s, and define p as:

$$p = m_s + l_s = \left\{ \Delta E/\hbar - \sum_k [r_k + \gamma_k(l_k - 4\bar{S}_k\bar{n}_k(\bar{n}_k + 1) + 2\bar{n}_k + 1)]\tilde{\omega}_k - \sum_{k\neq s}(m_k + l_k)\tilde{\omega}_k \right\} \Big/ \tilde{\omega}_s$$

$$= -\Delta G^\circ/\hbar\tilde{\omega}_s - \left\{ \sum_k [r_k + \gamma_k(l_k - 4\bar{S}_k\bar{n}_k(\bar{n}_k + 1))]\tilde{\omega}_k + \sum_{k\neq s}(m_k + l_k)\tilde{\omega}_k \right\} \Big/ \tilde{\omega}_s. \tag{A.136}$$

Following Kakitani & Kakitani (1981), $\Delta E - \sum_k(2\bar{n}_k + 1)\hbar\tilde{\omega}_k\gamma_k$ has been replaced by $-\Delta G^\circ$ which it equals in a first-order approximation.* Of

* In a system with only a vibrational degree of freedom, $G^\circ = E^\circ + k_B T \ln[\sinh(\hbar\omega/2k_B T)]$ in the harmonic approximation. From this one can derive

course, p may be defined in any way convenient for computation. The choice of $p = m_s + l_s$ rests on the fact that these two indices fuse into one as γ goes to 0. Equation (A.136) actually only calculates m_s since l_s has already been chosen arbitrarily in doing the sum over k.

With this definition for p, the approximate expression for the rate of electron transfer becomes:

$$W = \frac{2\pi}{\hbar^2}|H_{RP}|^2 \exp(-\bar{S}_s(2\bar{n}_s + 1)) \prod_{k \neq s} \exp(-\bar{S}_k(2\bar{n}_k + 1))$$
$$\times \left[\sum_{m_k=-\infty}^{\infty} (-1)^m J_{m_k}(4\bar{S}_k\bar{n}_k(\bar{n}_k + 1))\right]$$
$$\times \left[\sum_{l_k=-\infty}^{\infty} \left(\frac{\bar{n}_k + 1}{\bar{n}_k}\right)^{l_k/2} J_{l_k}(2\bar{S}_k(2\bar{n}_k + 1)(\bar{n}_k(\bar{n}_k + 1))^{\frac{1}{2}})\right]$$
$$\times \left[\sum_{r_k=-2}^{2} u(r_k)\right] \times u(r_s)\left(\frac{\bar{n}_s + 1}{\bar{n}_s}\right)^{l_s/2} J_{l_s}(2\bar{S}_s(2\bar{n}_s + 1)$$
$$\times (\bar{n}_s(\bar{n}_s + 1))^{\frac{1}{2}})(-1)^{p-l_s} J_{p-l_s}(4\bar{S}\bar{n}_s(\bar{n}_s + 1)) \qquad \text{(A.137)}$$

Lin (1966) derived formulas for the multiphonon process considering both displacement and frequency change of the oscillators. He used as his matrix element that given by equation (A.40) with the second term in the right-hand member of equation (A.41) substituted into it. He applies the Condon approximation to factor out $\phi_m \partial\phi_n/\partial Q_k$ and so uses in place of equation (5.21) the following (his equation (18) with our symbols):

$$W_{mn} = \frac{2\pi}{\hbar}\sum_k \frac{1}{2\pi\hbar}|R_k(mn)|^2 \int \mathrm{d}t\, \mathrm{e}^{-i\Delta Et/\hbar} K_k \prod_l G_l. \qquad \text{(A.138)}$$

The subscripts k and l refer to the different vibrational modes. $R_k(mn)$ is the factored out $-\hbar^2(M_k)^{-1/2}\langle \phi_m|\partial/\partial Q_k|\phi_n\rangle$. The factor K_k contains the derivatives of the Franck–Condon overlap integrals generated by $\partial\chi_{nj}/\partial Q_k$. The factors G_l contain the overlap integrals $\langle \chi_{mi}|\chi_{nj}\rangle$ (or $(C(i,j))$ (see equation (5.36)). I hope the reader is not confused by my use of n both to designate vibrational states as in equation (5.36) and everything following and also to designate electronic states as in equations (A.40) and (A.138) and when subscripted to ϕ. In (A.138) i and j designated vibrational states before and after electron transfer, respectively.

To compare Lin's results with ours, note that we have used an electronic matrix element, designated H_{RP}, which does not overtly generate derivatives of the overlap integrals. The correspondence between the electronic

---

the above approximation to $-\Delta G^\circ$. I use ΔG° to mean final – initial as in chemical thermodynamic usage while (perhaps unfortunately) I retain the common usage in multiphonon theory of ΔE = initial – final.

matrix elements is, then:

$$|H_{RP}|^2 \leftrightarrow \sum_k |R_k(mn)|^2 K_k. \tag{A.139}$$

I am not saying that these are equal, because K_k depends upon t and cannot properly be taken out of the integral over t but a factor with the dimensions of K_k is necessary to give the same dimensions on each side of (A.139) and in our formulation K_k has no place in the Franck–Condon factor.

The part of Lin's formulation equivalent to our factor (FC) as used in equation (5.21) and thereafter, is then:

$$(\mathrm{FC}) = \frac{1}{2\pi\hbar}\int \mathrm{d}t e^{-i\Delta Et/\hbar} \prod_l G_l. \tag{A.140}$$

It can be shown that Lin's equation (22), which gives his formula for G_l, corrected for misprints,* substituted into (A.140) gives precisely our equation (A.123) and, therefore, (A.124).

To proceed beyond these exact expressions, Lin makes approximations as follows: his equations (26) and (28) together have the same result as that of equation (A.125) and neglect of γ^2 in the pre-exponential part of (A.124). His equation (27) asserts the approximation of equation (A.126) and further entirely neglects γ as it appears in the coefficient of $\bar{S}$ in equation (A.128) and uses ω for $\tilde{\omega}$. This is compatible with his assumption that $\bar{S}$ is as small in order of magnitude as γ, but not with the non-small values of $\bar{S}$ that we find.

Kubo in his original paper (Kubo,1952) treated the case having frequency change with no displacement of equilibrium position of the oscillators.

A.7 Transition from state *a* to a set of states *b*

This section outlines the details of the full quantum-mechanical calculation of the problem presented in Fig. A.4, Section A.1.3(*a*), which leads to the results shown in Fig. A.5. The treatment is, in principle, the same as that of Bixon & Jortner (1968) (their equations (10)–(29)) with variations in detail.

Here I use the Dirac (1935) bra and ket notation to represent states of the system. The base-states shown in Fig. A.4 are then $|a\rangle, |b_{-N}\rangle, |b_{-N+1}\rangle, \cdots, |b_0\rangle, |b_1\rangle, \cdots, |b_N\rangle$. There are $2n + 2 = M$ states. The 'exchange' (or) 'configuration' interaction $|H_{ab}|$ between $|a\rangle$ and the set $\{|b\rangle\}$ is assumed to be equal to the constant, A, for all the b and will be used as the

* The third 'coth' in Lin's equation (22) should be 'tanh' and the coefficient of coth $\frac{1}{2}\lambda'_j$ in the exponent should be $\beta''j^2$ instead of $\beta'j^2$.

unit of energy. H_{aa} will be the zero of energy. The spacing between successive b states is d so that the density of b states is $\rho = (Ad)^{-1}$. The Hamiltonian matrix then has the form:

$$H_{ij} = \begin{array}{c|ccccccc} & a & b_N & b_{N+1} & \cdots & b_0 & \cdots & b_N \\ \hline a & 0 & 1 & 1 & \cdots & 1 & \cdots & 1 \\ b_N & 1 & -\mathrm{N}d & 0 & \cdots & 0 & \cdots & 0 \\ b_{N+1} & 1 & 0 & (-N+1)d & \cdots & 0 & \cdots & 0 \\ \vdots & \vdots & \vdots & \vdots & \vdots & \vdots & \vdots & \vdots \\ b_0 & 1 & 0 & 0 & \cdots & 0 & \cdots & . \\ \vdots & \vdots & \vdots & \vdots & \vdots & \vdots & \vdots & \vdots \\ b_N & 1 & 0 & 0 & \cdots & 0 & \cdots & Nd \end{array} \tag{A.141}$$

The states referred to by i and j are indicated at the heads of the rows and columns.

The eigenvectors of the Hamiltonian express eigenstates as expansions in the base-states:

$$|\psi_m\rangle = \sum_{k=a,\,-N}^{N} X_{mk}|a, b_k\rangle \qquad m = 1, 2, \ldots, M. \tag{A.142}$$

Symbolically $X_{mk} = \langle k|\psi_m\rangle$, which means 'the amplitude to be in state $|k\rangle$ given that the system is in state $|\psi_m\rangle$', and $|k\rangle$ stands for a base-state, $|a\rangle$ or $|b_k\rangle$. A general state of the system is expanded in terms of the eigenstates with their time dependences as follows:

$$|\Psi, t\rangle = \sum_{m=1}^{M} g_m|\psi_m\rangle e^{-iE_m t/\hbar}, \tag{A.143}$$

where E_m = the eigenvalue for eigenstate, $|\psi_m\rangle$ and, so far, g_m is arbitrary. However, we next require the system to be in state $|a\rangle$ at $t = 0$. Equation (A.143) then becomes:

$$|\Psi, 0\rangle = \sum_{m=1}^{M} g_m|\psi_m\rangle = |a\rangle. \tag{A.144}$$

This means that g_m is the amplitude $\langle\psi_m|a\rangle$, to be in $|\psi_m\rangle$ if the system is known to be in $|a\rangle$. By quantum-mechanical rules: $g_m = \langle\psi_m|a\rangle = \langle a|\psi_m\rangle^* = X_{ma}{}^*$, the complex conjugate of X_{ma}.

The amplitude to be in $|a\rangle$ at time t is:

$$\begin{aligned} C_a(t) = \langle a, |\Psi, t\rangle &= \sum_{m=1}^{M} g_m\langle a|\psi_m\rangle e^{-iE_m t/\hbar} \\ &= \sum_{m=1}^{M} X_{ma}{}^* X_{ma} e^{-iE_m t/h} \end{aligned} \tag{A.145}$$

Note that this is summing before 'squaring'. The 'squaring' comes next.

The probability to be in $|a\rangle$ at time t is:

$$C_a^{\,2}(t) = \left(\sum_{m=1}^{M} X_{ma}{}^{*}X_{ma}\mathrm{e}^{-iE_m t/\hbar}\right)\left(\sum_{l=1}^{M} X_{la}X_{la}{}^{*}\mathrm{e}^{iE_l t/\hbar}\right). \qquad \text{(A.146)}$$

The cross-products occurring in equation (A.146) can be grouped as follows:

if $m = l$, product $= (X_{ma}{}^{*}X_{ma})^2$;

if $m \neq l$, product $= X_{ma}{}^{*}X_{ma}X_{la}X_{la}{}^{*}\mathrm{e}^{i(E_l - E_m)t/\hbar}$

The latter can be grouped into pairs chosen symmetrically about the diagonal of an m, l matrix:

$$\begin{aligned}\text{sym. pair: } & X_{ma}{}^{*}X_{ma}X_{la}X_{la}{}^{*}[E^{i(E_l - E_m)t/\hbar} + \mathrm{e}^{-i(E - E_m)t/\hbar}] \\ & = 2X_{ma}{}^{*}X_{ma}X_{la}X_{la}{}^{*}\cos[(E_l - E_m)t/\hbar].\end{aligned}$$

This enables equation (A.146) to be written:

$$\begin{aligned}C_a^{\,2} = {} & \sum_{m=1}^{M} (X_{ma}{}^{*}X_{ma})^2 \\ & + 2\sum_{m=1}^{M-1} X_{ma}{}^{*}X_{ma}\sum_{l=m+1}^{M} X_{la}X_{la}{}^{*}\cos[(E_l - E_m)t/\hbar]. \qquad \text{(A.147)}\end{aligned}$$

To solve for the eigenvalues one subtracts E from each of the diagonal elements of (A.141), makes a determinant from the result, equates it to zero and solves for E. This can be written:

$$E\prod_{j=-N}^{N}(jd - E) - \sum_{i=-N}^{N}\left[\prod_{l=-N}^{i-1}\prod_{l=i+1}^{N}(ld - E)\right] = 0 \qquad \text{(A.148)}$$

Equation (A.148) which is solved numerically by trial and error avoids the possibility of requiring division by zero. A similar equation can be made by factoring out $\prod_{j=-N}^{N}(jd - E)$ and is suggested by Bixon & Jortner (1968, their equation (15)), but it requires division by zero for certain trial values of E (but not for any eigenvalue).

The ratios, r_{mk}, of the eigenvector components, X_{mk}/X_{ma}, corresponding to eigenvalue E_m are found by solving the simultaneous equations:

$$\sum_{j=-N}^{N} H_{ij}r_{mj} = E_m r_{mi}, \quad i = -N, \ldots, N. \qquad \text{(A.149)}$$

In the ranges indicated, however, $H_{ij} = id\delta_{ij}$, so that

$$r_{mi} = (E_m - id)^{-1}. \qquad \text{(A.150)}$$

Normalization requires that:

$$X_{ma} = \left(1 + \sum_{i=-N}^{N} r_{mi}{}^2\right)^{-1/2}. \qquad \text{(A.151)}$$

Equations (A.150) and (A.151) are substituted with the E_m from equation (A.148) into equation (A.147) to obtain $C_a^{\,2}(t)$. The occupation of $\{|b\rangle\}$ is then:

$$C_b^{\,2}(t) = 1 - C_a^{\,2}(t). \tag{A.152}$$

Note that the X_{ma} are all real so that $X_{ma} = X_{ma}{}^*$.

The work of Bixon & Jortner (1968) differs from the above in part because they do not confine themselves to a finite number of states. They arrive at exponential behaviour with the golden rule time-constant as above. They specify two validity criteria. One, that (in our notation) $d \ll |H_{ab}|$, is the same as the criterion $A\rho \gg 1$, expressed here in Section A.1.3 (*a*). The other is that $t \ll \hbar/d$. Since the times of interest are of the order of the decay time constant, $\tau = \hbar/2\pi|H_{ab}|^2\rho$, we might substitute this expression for t and obtain $|H_{ab}|\rho \gg (2\pi)^{-1/2}$, which is practically the same as the first criterion. Our second criterion, that $N/(A\rho)^2 \gg 1$, deals with the finite number of states used in the numerical calculation and is not pertinent if N is taken to be infinite.

References

Throughout this list the abbreviation **T** stands for the reference: Chance *et al.*, eds. (1979*a*), *q.v.*

Ahmed, N. A. G., Calderwood, J. H., Frolich, H. & Smith, C. W. (1975). Evidence for the collective magnetic effects in an enzyme. Likelihood of room temperature superconductive regions. *Phys. Lett.* **35A**, 129–30.

Ahmed, N. A. G., Smith, C. W., Calderwood, J. H. & Frolich, H. (1976). Electric and magnetic properties of lysozyme and other biomolecules. *Collective Phen.* **2**, 155–66.

Alben, J. O., Beece, D., Bowne, S. F., Eisenstein, L., Frauenfelder, H., Good, D., Marden, M. C., Moh, P. P., Reinisch, L., Reynolds, A. H. & Yue, K. T. (1980). Isotope effect in molecular tunneling. *Phys. Rev. Lett.* **44**, 1157–60.

Alberding, N., Austin, R. H., Beeson, K. W., Chan, S. S., Eisenstein, L., Frauenfelder, H. & Nordlund, T. M. (1976). Tunneling in ligand binding to heme proteins. *Science NY* **192**, 1002–3.

Alkaitis, S. A., Grätzel, M. & Henglein, A. (1975). Laser photoionization of phenothiazine in micellar solution. II. Mechanism and light induced redox reactions with quinones. *Ber. Bunsenges. phys. Chem.* **79**, 541–6.

Allen, G. C. & Hush, N. S. (1967). Intervalence-transfer absorption. 1. Qualitative evidence for intervalence-transfer absorption in inorganic systems in solution and in the solid state. *Prog. Inorg. Chem.* **8**, 357–89.

Anderson, P. W. (1950). Antiferromagnetism. Theory of superexchange interaction. *Phys. Rev.* **79**, 350–6.

Anderson, P. W. & Rowell, J. M. (1963). Probable observation of the Josephson superconducting tunneling effect. *Phys. Rev. Lett.* **10**, 230–2.

Armstrong, C. M. (1974). Ionic pores, gates, and gating currents. *Quart. Rev. Biophys.* **7**, 179–209.

Arnold, W. & Clayton, R. K. (1960). The first step in photosynthesis: evidence for its electronic nature. *Proc. natn. Acad. Sci. USA* **46**, 769–76.

Austin, R. H., Beeson, K. W., Eisenstein, L., Frauenfelder, H. & Gunsalus, I. C. (1975). Dynamics of ligand binding to myoglobin. *Biochemistry* **14**, 5355–573.

Austin, R. H., Beeson, K. W., Eisenstein, L., Frauenfelder, H., Gunsalus, I. C. & Marshall, V. P. (1974). Activation energy spectrum of a biomolecule: photodissociation of carbonmonoxy myoglobin at low temperatures. *Phys. Rev. Lett.* **32**, 403–5.

Austin, R. H. & Hopfield, J. J. (1982). Attempted time-resolved measurement of photo-assisted charge transfer of cytochrome c iron hexacyanide. In *Electron Transport and Oxygen Utilization* (*Arlie Symp. 4*) (ed. C. Ho, W. A. Eaton, J. P. Collman, Q. H. Gibson, J. S. Leigh, Jr, E. Margoliash, J. K. Moffet and W. R. Scheidt), pp. 73–80. Amsterdam: Elsevier/North Holland.

Avouris, P., Peters, K. S. & Rentzepis, P. M. (1978). Direct measure of electron transfer in bacteriochlorophyll. *Biophys. J.* **21**, 8a. Abstr M-Am-Al.

Azumi, T. & Matsuzaki, K. (1977). What does the term 'vibronic coupling' mean? *Photochem. Photobiol.* **25**, 315–26.

Ballard, S. G. & Mauzerall, D. (1979). Kinetic, evidence for electron tunneling in solution. **T**, pp. 581–7.

Ballard, S. G. & Mauzerall, D. (1980). Photochemical ionogenesis in solutions of zinc octaethyl porphyrin. *J. chem. Phys.* **72**, 933–47.

Barber, J. (ed.) (1977). *Primary Processes of Photosynthesis.* Amsterdam: Elsevier/North Holland.

Bardeen, J. (1961). Tunneling from a many-particle point of view. *Phys. Rev. Lett.* **6**, 57–9.

Bardeen, J., Cooper, L. N. & Schrieffer, J. R. (1957). Theory of superconductivity. *Phys. Rev.* **108**, 1175–204.

Barkalov, I. M., Gareeva, D. A., Goldanskii, V. I., Enikolopyan, N. S. & Berlin, Al.Al. (1966). *Vysokomol Soed.* **8**, 1140. (*English Trans.:* Radiation polymerization of monomers irradiated at 4.2 K. *Polymer Sci. USSR* **8**, 1255–6.)

Barkalov, I. M., Goldanskii, V. I., Kiryukhin, D. P. & Zanin, A. M. (1980). Kinetics and mechanism of the low-temperature chain hydrobromination of eythlene. *Chem. Phys. Lett.* **73**, 273–7.

Bartsch, R. G. & Kamen, M. D. (1960). Isolation and properties of two soluble heme proteins in extracts of the photoanaerobe *Chromatium. J. biol. Chem.* **235**, 825–31.

Bassham, J. A., Benson, A. A., Kay, L. D., Harris, A. Z., Wilson, A. T. & Calvin, M. (1954). The path of carbon in photosynthesis. XXI. The cyclic regeneration of carbon dioxide acceptor. *J. Am. chem. Soc.* **76**, 1760–70.

Bateman, H. (1953). *Higher Transcendental Functions*, vol. 11 (ed. A. Erdélyi *et al.*). New York: McGraw-Hill.

Bawn, C. E. H. & Ogden, G. (1934). Wave mechanical effects and the reactivity of the hydrogen isotopes. *Trans. Faraday Soc.* **30**, 432–43.

Beitz, J. V. & Miller, J. R. (1979*a*). Tunneling distances and exothermic rate restrictions in electron transfer reactions. **T**, pp. 269–80.

Beitz, J. V. & Miller, J. R. (1979*b*). Exothermic rate restrictions on electron transfer in a rigid medium. *J. chem. Phys.* **71**, 4579–95.

Bell, R. P. (1933). The application of quantum mechanics to chemical kinetics. *Proc. R. Soc. Lond.* A **139**, 466–74.

Bell, R. P. (1935). Quantum mechanical effects in reactions involving hydrogen. *Proc. R. Soc. Lond.* A **148**, 241–50.

Benderskii, V. A., Goldanskii, V. I. & Ovchinnikov, A. A. (1980). Effect of molecular motion on low-temperature and other anomalously fast chemical reactions in the solid phase. *Chem. Phys. Lett.* **73**, 492–5.

Birks, J. B. (1959). General discussion. *Discussion Faraday Soc.* **27**, 243–4.

Bixon, M. & Jortner, J. (1968). Intramolecular radiationless transitions. *J. chem Phys.* **48**, 715–26.

Blankenship, R. E. & Parson, W. W. (1979). Kinetics and thermodynamics of electron transfer in bacterial reaction centers. In *Photosynthesis in Relation to Model Systems* (ed. J. Barber), pp. 71–114. Amsterdam: Elsevier/North Holland.

Blankenship, R. E., Schaafsma, T. J. & Parson, W. W. (1977). Magnetic field effects on radical pair intermediates in bacterial photosynthesis. *Biochim. biophys. Acta* **461**, 297–305.

Blois, M. S. Jr (1971). Physical studies of the melanins. In *Biology of Normal and Abnormal Melanocytes* (ed. T. Kawamura, T. B. Fitzpatric and M. Seiji), pp. 125–40. Baltimore: University Park.

Blumenfeld, L. A. (1978). The physical aspects of energy transduction in biological system. *Q. Rev. Biophys.* **11**, 251–308.

Blumenfeld, L. A. & Chernavskii, D. S. (1973). Tunneling of electrons in biological processes. *J. theor. Biol.* **39**, 1–7.

Boardman, N. K. & Anderson, J. M. (1964). Isolation from spinach chloroplasts of particles containing different proportions of chlorophyll *a* and chlorophyll *b* and their possible role in the light reactions of photosynthesis. *Nature, Lond.* **203**, 166–167.

Bogomolni, R. A. & Klein, M. P. (1975). Mobile charge carriers in photosynthesis. *Nature, Lond.* **258**, 88–9.

Bogomolni, R. A. & Klein, M. P. (1979). Faraday rotation and photoconductivity of photosynthetic structures at microwave frequencies. **T**, pp. 405–16.

Bone, S., Lewis, T. J., Pethig, R. & Szent-Gyorgyi, A. (1978). Electronic properties of some protein-methylglyoxal complexes. *Proc. natn. Acad. Sci. USA* **75**, 315–18.

Born, M. & Huang, K. (1954). *Dynamical Theory of Crystal Lattice.* New York: Oxford University Press.

Born, M. & Oppenheimer, J. R. (1927). Zur Quantentheorie der Molekeln. *Annln Phys.* **84**, 457–84.

Bourgin, D. G. (1929). Unimolecular reactions. *Proc. natn. Acad. Sci. USA* **15**, 357–62.

Brailsford, A. D. & Chang, T. Y. (1970). Nonradiative decay of individual vibronic levels in large molecules. *J. chem. Phys.* **53**, 3108–13.

Brill, A. S. (1978). Activation of electron transfer reactions of the blue proteins. *Biophys. J.* **22**, 139–42.

Brill, A. S. (1979). Conformational distribution and vibronic coupling in the blue copper-containing protein azurin. **T**, pp. 561–8.

Brillouin, L. (1926). La mécanique ondulatoire de Schrödinger; une générale de résolution par approximations successives. *Comptes Rendus* **183**, 24–6.

Brocklehurst, B. (1973). An electron-tunneling model for recombination of aromatic hydrocarbon radical ions in non-polar solvents. *Chem. Phys.* **2**, 6–18.

Brocklehurst, B. (1979). Electron tunneling after radiolysis. **T**, pp. 243–67.

Brocklehurst, B., Bull, D. C. & Evans, M. (1975). Thermoluminescence of solutions of squalane after γ-irradiation. *J. Chem. Soc., Far. Trans. II*, **71**, 543.

Brönsted, J. N. (1928). Acid and basic catalysis. *Chem. Rev.* **5**, 231–338.

Bruk, M. A., Gromov, V. F., Chernyak, I. V., Khomikovskii, P. M. & Abkim, A. D. (1966). *Vysokomol. Soed.* **8**, 961–2. (*English Trans.:* Radiation polymerization of tetrafluoroethylene and acrylonitrile at 4.2 K. *Polymer Sci. USSR* **8**, 1059–60.)

Brunschwig, B. S., Logan, J., Newton, M. D. & Sutin, N. (1980). A semiclassical treatment of electron-exchange reactions. Application to the hexaaquoiron (II)–hexaaquoiron (III) system. *J. Am. chem. Soc.* **102**, 5798–809.

Dudgor, A. (1979). Tunneling in network models of molecular chains. **T**, pp. 77–90.

Buhks, E., Bixon, M. & Jortner, J. (1981). Deuterium isotope effects on outersphere electron-transfer reactions. *J. Phys. Chem.* **85**, 3763–3766.

Bunker, B. C., Drago, R. S., Hendrickson, D. N., Richman, R. M. & Kessell, S. L. (1978). Experimental evidence for trapped valences in a mixed-valence complex μ-pyrazine-bis (penta-ammineruthenium) tosylate. Electron paramagnetic resonance, magnetic susceptibility, and nuclear magnetic resonance results. *J. Am. chem. Soc.* **100**, 3805–14.

Burrau, Ø. (1927). Berechnung des Energiewertes des Wasserstoff molekelions, (H_2^+) im Normalzustand. *Det. Kgl. Danske Vid. Selskab.* **7**, nr. 14, 1–18.

Burstein, E. & Lundqvist, S. (eds.) (1969). *Tunneling Phemenon in Solids.* New York: Plenum.

Busch, H. (1962). *Biochemistry of the Cancer Cell* (especially pp. 202–7). New York: Academic Press.

Butler, J., Jayson, G. G. & Swallow, A. J. (1975). The reaction between the superoxide anion radical and cytochrome *c*. *Biochim. biophys. Acta* **408**, 215–22.

Butler, W. L. (1973). Primary photochemistry of photosystem II of photosynthesis. *Acc. Chem. Res.* **6**, 117–84.

Butler, W. L., Visser, J. W. M. & Simons, H. L. (1973). The kinetics of light-induced changes of C-550, cytochrome *b* 559 and fluorescence yield in chloroplasts at low temperature, *Biochem. biophys. Acta* **292**, 140–51.

Buxton, G. V. & Kemsley, K. G. (1976). Low temperature pulse radiolysis of concentrated aqueous solutions. *J. chem. Soc. Far. Trans. I*, **72**, 466–80.

Calvin, M. (1956). The photosynthetic carbon cycle. *J. chem. Soc. pp.* 1895–915.

Calvin, M. & Sogo, P. B. (1975). Primary quantum conversion process in photosynthesis: electron spin resonance. *Science, N.Y.* **125**, 499–500.

Cartling, B. & Ehrenberg, A. (1979). The energy transforming function associated with electron transfer reactions in biological systems. **T**, pp. 575–80.

Chance, B. (1979). (General discussion on biological specificity and control). **T**, pp. 609–10.

Chance, B. & Bonner, W. D., Jr (1963). The temperature insensitive oxidation of cytochrome F in green leaves – a primary biochemical event of photosynthesis. In *Photosynthesis Mechanisms in Green Plants*, pp. 66–81. National Acad. Science–National Research Council Publication, no. 1145.

Chance, B., DeVault, D., Frauenfelder, H., Marcus, R. A., Schrieffer, J. R. & Sutin, N. (eds.) (1979*a*). *Tunneling in Biological Systems.* New York: Academic Press. This reference is abbreviated **T** in refernces to its chapters.

Chance, B., DeVault, D., Tasaki, A. & Thornber, J. P. (1979*b*). The effects of high hydrostatic pressure on light-induced electron transfer and proton binding in Chromatium. **T**, pp. 387–402.

Chance, B. & Hollunger, G. (1957). Succinate-linked pyridine nucleotide reduction in mitochondria. *Fdn. Proc.* **16**, Abstr. 703, p. 163.

Chance, B. & Hollunger, G. (1961). The interaction of energy and electron transfer reactions in mitochondria. I. General properties and nature of the products of succinate-linked reduction of pyridine nucleotide. *J. biol. Chem.* **236**, 1534–43.

Chance, B., Kihara, T., DeVault, D., Hildreth, W., Nishimura, M. & Hiyama, T. (1969). Temperature insensitive electron transfer in photosynthetic systems. In *Progress in Photosynthesis Research* (ed. H. Metzner), pp. 1321–46. Tübingen.

Chance, B. & Nishimura, M. (1960). On the mechanism of chlorophyll–cytochrome interaction: the temperature insensitivity of light-induced cytochrome oxidation in chromatium. *Proc. natn. Acad. Sci. USA* **46**, 19–24.

Chance, B., Saronio, C. & Leigh, J. S., Jr (1975). Functional intermediate in reaction of cytochrome oxidase and oxygen. *Proc. natn. Acad. Sci. USA* **72**, 1635–40.

Chance, B., Saronio, C., Leigh, J. S., Jr & Waring, A. (1979*c*). Electron transfer reactions in cytochrome oxidase. **T**, pp. 483–511.

Chance, B. & Williams, G. R. (1956). The respiratory chain and oxidative phosphorylation. In *Advances in Enzymology* (ed. F. F. Nord), pp. 65–134. New York: Interscience.

Chang, A. & Austin, R. H. (1982). Electron tunneling in cytochrome *c*. *J. Chem. Phys.* **77**, 5272–83.

Chen, E. C. M. & Wentworth, W. E. (1975). A comparison of experimental determinations of electron affinities of PI charge transfer complex acceptors. *J. chem. Phys.* **63**, 3183–94.

Chernavskaya, N. M. & Chernavskii, D. S. (1978). Fizicheskie aspekti v funktsionirovanii bakteriorodapsina. *Academy of Sciences, USSR, Physics Institute*, Preprint no. 61. Moscow.

Chernavskaya, N. M., Chernavskii, D. S., Günther, K., Hache, A. & Kilyachkov, A. A. (1977). On the mechanism of energy transformation in photosynthesis. *Stud. Biophys.* **62**, 109–26.

Child, M. S. (1967). Measurable consequences of a dip in the activation barrier for an adiabatic chemical reaction. *Molec. Phys.* **12**, 401–16.

Chou, M., Creutz, C. & Sutin, N. (1977). Rate constants and activation parameters for outer-sphere electron transfer reactions and comparisons with the predictions of Marcus theory. *J. Am. chem. Soc.* **99**, 5615–23.

Christov, St G. (1958). Die Rolle des Tunnelübergangs der Ionen in der Kinetik der Elektrodenvorgänge. *Z. Elektrochem.* **62**, 567–81.

Christov, St G. (1960). Die anwendung der parabolischen Potentialschwelle auf die Untersuchung der Quanteneffekte in der Elektrochemie. *Z. Elektrochem.* **64**, 840–8.

Christov, St G. (1975). Quantum theory of electron-transfer processes in solution. *Ber. Bunsenges. phys. Chem.* **79**, 357–71.

Clarke, J. (1974). Josephson junction detectors. *Science, NY* **184**, 1234–42.

Clayton, R. K. (1979). Some data of possible relevance to tunneling in photosynthetic reaction centers. **T**, pp. 377–86.

Clayton, R. K. & DeVault, D. (1972). Effects of high pressure on photochemical reaction centers from *Rhodopseudomonas spheroides. Photochem. & Photobiol.* **15**, 165–75.

Clayton, R. K. & Wang, R. T. (1971). Photochemical reaction centres from *Rhodopseudomonas spheroides*. In *Methods in Enzymology*, **23** (ed. S. P. Colwick, N. Kaplan and A. San Pietro), pp. 696–704. New York: Academic Press.

Clayton, R. K. & Yau, H. F. (1972). Photochemical electron transport in photosynthetic reaction centres from *Rhodopseudomonas spheroides*. I. Kinetics of the oxidation and reduction of P-870 as affected by external factors. *Biophys. J.* **12**, 867–81.

Cohen, M. H., Falicov, L. M. & Phillips, J. C. (1962). Superconductive tunneling. *Phys. Rev. Lett.* **8**, 316–18.

Coleman, R. V. (1979). Detection of molecular vibrations with electron tunneling. **T**, pp. 67–76.

Condon, E. U. (1926). A theory of intensity distribution in band systems. *Phys. Rev.* **28**, 1182–201.

Condon, E. U. (1928). Nuclear motions associated with electron transitions in diatomic molecules. *Phys. Rev.* **32**, 858–72.

Condon, E. U. & Morse, P. M. (1929). *Quantum Mechanics.* New York: McGraw-Hill.

Cope, F. W. (1965*a*). A generalized theory of particulate electron conduction enzymes applied to cytochrome oxidase. A theory of coupled electron and/or ion transport applied to pyruvate carboxylase. *Bull. math. Biophys.* **27**, 237–52.

Cope, F. W. (1965*b*). A kinetic theory of electron and ion transport in particulate and membraneous systems, with applications to cytochrome oxidase, melanin free radical and pyruvate carboyxylase reactions, and to control of enzymes by hormones and radiation. In *Oxidases and Related Redox Systems* (ed. T. King, H. Mason and M. Morrison), pp. 51–71. New York: Wiley.

Cope, F. W. & Straub, K. D. (1969). Calculation and measurement of semiconduction activation energy and electron mobility in cytochrome oxidase, with evidence that charge carriers are polarons, which may couple oxidation to phosphorylation. *Bull. math. Biophys.* **31**, 761–72.

Cramer, W. A. & Whitmarsh, J. (1979). An application of electron transfer theory to a problem in chloroplast membrane topography. **T**, pp. 363–70.

Cramer, W. A., Whitmarsh, J. & Horton, P. (1979). Cytochrome *b* in energy-transducing membranes. In *The Porphyrins*, vol. vii (ed. D. Dolphin), pp. 71–106. New York: Academic Press.

Creutz, C. & Sutin, N. (1977). Vestiges of the 'inverted region' for highly exergonic electron-transfer reactions. *J. Am. chem. Soc.* **99**, 241–3.

Cribb, P. H., Nordholm, S. & Hush, N. S. (1979). A general theoretical approach to tunneling transfer and dissociation. **T**, pp. 139–44.

Cusanovich, M. A. & Bartsch, R. G. (1969). A high potential cytochrome *c* from *Chromatium* chromatophores. *Biochim. biophys. Acta* **189**, 245–55.

Cusanovich, M. A. & Mizrahi, I. A. (1979). Electron transfer between *c*-type cytochromes and high potential iron–sulfer proteins. **T**, pp. 355–62.

Davydov, A. S. (1982). *Biology and Quantum Mechanics.* Oxford: Pergamon.

Desouter-Lecomte, M. & Lorquet, J. C. (1977). Nonadiabatic interactions in unimolecular decay. II. Simplified formalism. *J. Chem. Phys.* **66**, 4006–17.

Desouter-Lecomte, M. & Lorquet, J. C. (1979). Nonadiabatic interactions in unimolecular decay. IV. Transition probability as a function of the Massey parameter. *J. chem. Phys.* **71**, 4391–403.

DeVault, D. (1979). Introduction to biological aspects. **T**, pp. 303–16.

DeVault, D. (1980). Quantum mechanical tunnelling in biological systems. *Quart. Rev. Biophys.* **13**, 387–564.

DeVault, D. & Chance, B. (1966). Studies of photosynthesis using a pulsed laser. I. Temperature dependence of cytochrome oxidation rate in *chromatium*. Evidence for tunneling. *Biophys. J.* **6**, 825–47.

DeVault, D., Parkes, J. H. & Chance, B. (1967). Electron tunneling in cytochromes. *Nature, Lond.* **215**, 642–4.

DeVault, D. & Sarai, A. (1983). A comparison of matrix elements for electron transfer. *Biophys. J.* **41**, Abstract T-PM-Pos 35.

Dexter, D. L. (1953). A theory of sensitized luminescence in solids. *J. chem. Phys.* **21**, 836–50.

Dirac, P. A. M. (1927). The physical interpretation of the quantum dynamics. *Proc. R. Soc. Lond.* A **113**, 621–41.

Dirac, P. A. M. (1935). *Quantum Mechanics.* New York: Oxford University Press.

Dogonadze, R. R., Kharkats, Yu. I. & Ulstrup, J. (1976). A dynamic quantum mechanical model for the elementary act of spontaneous and induced mutations. *Chem. Phys. Lett.* **37**, 360–4.

Dogonadze, R. R., Kuznetsov, A. & Chernenko, A. (1965). Teoriya gomogennikh i geterogennikh elektronikh protsessov v shidkostyakh. *Usp. Khim.* **34**, 1779–812.

Dogonadze, R. R., Kuznetsov, A. M. & Vorotyntsev, M. A. (1972*a*). On the theory of nonradiative transitions in polar media. I. Processes without 'mixing' of quantum and classical degress of freedom. *Phys. Status Solidi* B **54**, 125–134.

Dogonadze, R. R., Kuznetsov, A. M. & Vorotyntsev, M. A. (1972*b*). II. Processes with 'mixing' of quantum and classical degrees of freedom. *Phys. Status Solidi* B **54**, 425–33.

Dogonadze, R. R., Kuznetsov, A. M., Zakaraya, M. G. & Ulstrup, J. (1979). A quantum theory of low-temperature chemical and biological rate processes in condensed media. **T**, pp. 145–71.

Dogonadze, R. R., Ulstrup, J. & Kharkats, Yu. I. (1973*a*). A theory of polar medium electron transfer reactions through bridge groups with a quasicontinuous energy spectrum. *J. theor. Biol.* **40**, 259–77.

Dogonadze, R. R., Ulstrup, J. & Kharkats, Yu. (1973*b*). A theory of bridge-assisted polar medium electron transfer reactions between molecules having quasi continuous electronic energy spectra. *J. theor. Biol.* **40**, 279–83.

Douglass, D. H., Jr (1961). Direct experimental measurement of the magnetic field dependence of the superconducting energy gap of aluminum. *Phys. Rev. Lett.* **7**, 14–16.

Douzou, P. (1977). *Cryobiochemistry*. New York: Academic Press.

Duke, C. B. (1969). *Tunneling in Solids*. New York: Academic Press.

Duke, C. B. (1979). Concepts in quantum mechanical tunneling in systems of biological and chemical interest. **T**, pp. 31–65.

Dunham, J. L. (1932). The Wentzel–Brillouin–Kramers methods of solving the wave equation. *Phys. Rev.* **41**, 713–20.

Dutton, P. L. (1971). Oxidation-reduction potential dependence of the interaction of cytochromes, bacteriochlorophyll and carotenoids at 77 °K in chromatophores of *Chromatium D* and *Rhodopseudomonas gelationosa*. *Biochim. biophys. Acta* **226**, 63–80.

Dutton, P. L., Kaufmann, K. J., Chance, B. & Rentzepis, P. M. (1975). Picosecond kinetics of the 1250 nm band of the *Rps. sphaeroides* reaction center: the nature of the primary photochemical intermediary state. *FEBS Lett.* **60**, 275–80.

Dutton, P. L., Kihara, T., McCray, J. A. & Thornber, J. P. (1971). Cytochrome C_{553} and bacteriochlorophyll interaction at 77 °K in chromatophores and a subchromatophore preparation from Chromatium D. *Biochim. biophys. Acta* **226**, 81–7.

Dutton, P. L., Leigh, J. S., Jr, Prince, R. C. & Tiede, D. M. (1979). Cytochrome reaction center–quinone interactions: models for biological electron transfer. **T**, pp. 319–52.

Dutton, P. L., Leigh, J. S. & Seibert, M. (1972). Primary processes in photosynthesis: *in situ* ESR studies on the light induced oxidized and triplet state of reaction center bacteriochlorophyll. *Biochem. biophys. Res. Commun.* **46**, 406–13.

Dutton, P. L., Prince, R. C., Tiede, D. M., Petty, K. M., Kaufmann, K. J., Netzel, T. L. & Rentzepis, P. M. (1976). Electron transfer in the photosynthetic reaction center. In *Chlorophyll-proteins, Reaction Centers and Photosynthetic Membranes* (ed. J. M. Olson and G. H. Hind), pp. 213–36. Brookhaven Symposia in Biology, no. 28. Brookhaven National Laboratory, New York.

Dynes, R. C. (1979). Tunneling in physical systems. **T**, pp. 17–28.

Eckart, K. (1930). The penetration of a potential barrier by electrons. *Phys. Rev.* **35**, 1303–9.

Eigen, M. (1954). Über die Kinetik sehr schnell verlaufender Ionenreaktionen in wässeriger Lösung. *Z. phys. Chem.* (NF) **1**, 176–200.

Eley, D. D. & Leslie, R. B. (1964). Electrical conduction in solid proteins. In *Electronic Aspects of Biochemistry* (ed. B. Pullman), pp. 105–17. New York: Academic Press.

Eley, D. D. & Leslie, R. B. (1966). Kinetics of adsorption of water vapor and electrical conduction in bovine plasma albumin. *Trans. Faraday Soc.* **62**, 1002–14.

Eley, D. D. & Parfitt, G. D. (1955). The semiconductivity of organic substances, part 2. *Trans. Faraday Soc.* **51**, 1529–39.

Eley, D. D., Parfitt, G. D., Perry, M. H. & Taysum, D. H. (1953). The semiconductivity of organic substances, part 1. *Trans. Faraday Soc.* **49**, 79–86.

Eley, D. D. & Pethig, R. (1971). Microwave Hall effect measurements on organic and biological semiconductors. In *Conduction in Low-Mobility Materials* (ed. N. Klein, D. S. Tannhauser and M. Pollak) pp. 397–402. London: Taylor & Francis.

Eley, D. D. & Willis, M. R. (1961). The electrical conductivity of solid free radicals and the electron tunneling mechanism. In *Symposium on Electrical Conductivity in Organic Solids* (ed. H. Kallman and M. Silver), pp. 257–76. New York: Wiley.

Emin, D. (1975). Phonon-assisted transition rates. I. Optical phonon-assisted hopping in solids. *Adv. Phys.* **24**, 305–48.

Englman, R. & Jortner, J. (1970). The energy gap law for radiationless transitions in large molecules. *Molec. Phys.* **18**, 145–64.

Erecinska, M. (1979). Mitochondrial electron transfer at phosphorylation sites 2 and 3. **T**, pp. 453–70.

Erecinska, M., Blasie, J. K. & Wilson, D. F. (1977). Orientation of the hemes of cytochrome *c* oxidase and cytochrome *c* in mitochondria. *FEBS Lett.* **76**, 235–9.

Erecinska, M. & Chance, B. (1972). Studies on the electron transport chain at subzero temperatures: electron transport at site III. *Arch. Biochem. Biophys.* **151**, 304–15.

Erecinska, M., Wilson, D. F. & Blasie, J. K. (1978). Studies on the orientations of the mitochondrial redox carriers. I. Orientation of the hemes of cytochrome *c* oxidase with respect to the plane of a cytochrome oxidase-lipid model membrane. *Biochim. biophys. Acta* **501**, 53–62.

Esaki, L. (1958). New phenomenon in narrow germanium *p–n* junctions. *Phys. Rev.* **109**, 603–4. (L)

Esaki, L. (1974). Long journey into tunneling. *Science, NY* **183**, 1149–55.

Esaki, L. & Miyahara, Y. (1960). New device using the tunneling process in narrow p–n junctions. *Solis-St. Electron* **1**, 13–21.

Evans, M. C. W., Reeves, S. G. & Cammack, R. (1974). Determination of the oxidation-reduction potential of the bound iron–sulphur proteins of the primary electron acceptor complex of photosystem I in spinach chloroplasts. *FEBS Lett.* **49**, 111–14.

Evans, M. C. W., Sihra, C. K. & Cammack, R. (1976). The properties of the primary electron acceptor in photosystem I reaction centre of spinach chloroplasts and its interaction with P700 and the bound ferredoxin in various oxidation-reduction states. *Biochem. J.* **158**, 71–7.

Ewall, R. X. & Bennett, L. E. (1974). Reactivity characteristics of cytochrome *c* (III) adduced from its reduction by hexammineruthenium (II) ion. *J. Am. chem. Soc.* **96**, 940–2.

Eyring, H. (1935*a*). The activated complex in chemical reactions. *J. chem. Phys.* **3**, 107–15.

Eyring, H. (1935*b*). The activated complex and the absolute rate of chemical reactions. *Chem. Rev.* **17**, 65–77.

Fawcett, A. H. (1975). Formaldehyde polymers in interstellar space. *Nature. Lond.* **257**, 159.

Feher, G. (1971). Some chemical and physical properties of a bacterial reaction center particle and its primary photochemical reactants. *Photochem & Photobiol.* **14**, 373–87.

Feinberg, B. A., Aprahamian, G., Neeck, L., Ryan, M. D. & Wei, J-F. (1979). Overall charge control of the ionic strength effects upon the redox kinetics of small molecule–protein and protein–protein reactions. **T**, pp. 569–74.

Feinberg, B. A., Ryan, M. D. & Wei, J. F. (1977). Comparative kinetic-ionic strength study of two differently charged cytochromes *c*: effects are limited to overall charge. *Biochem. biophys. Res. Commun.* **79**, 769–75.

Ferguson-Miller, S., Brautigan, D. L. & Margoliash, E. (1978). Definition of cytochrome *c* binding domains by chemical modification. III. Kinetics of reaction of carboxydinitrophenyl cytochromes *c* with cytochrome *c* oxidase. *J. biol. Chem.* **253**, 149–59.

Ferguson-Miller, S., Brautigan, D. L. & Margoliash, E. (1979). Definition of the surface of cythochrome *c* interacting with cytochrome oxidase. **T**, pp. 513–21.

Feynman, R. P., Leighton, R. B. & Sands, M. (1965). *The Feynman lectures on physics.* Vol. III. *Quantum Mechanics.* New York: Addison-Wesley.

Filatovs, J., McGinness, J. & Corry, P. (1976). Thermal and electronic contributions to switching in melanins. *Biopolymers* **15**, 2309–12.

Fischer, H., Tom, G. M. & Taube, H. (1976). Intramolecular electron transfer mediated by 4,4′-bipyridine and related bridging groups. *J. Am. chem. Soc.* **98**, 5512–27.

Fischer, S. F. & Van Duyne, R. P. (1977). On the theory of electron transfer reactions. The naphthalene/TCNQ system. *Chem. Phys.* **26**, 9–16.

Fiske, M. D. (1964). Temperature and magnetic field dependences of the Josephson tunneling current. *Rev. mod. Physics* **36**, 221–2.

Floyd, R. A., Bronsdon, A. & Commoner, B. (1973). ESR signals during X-irradiation of tissue: their characteristics and relationship to the cancerous state. *Ann. NY Acad. Sci.* **222**, 1077–86.

Floyd, R. A., Chance, B. & DeVault, D. (1971). Low temperature photoinduced reactions in green leaves and chloroplasts. *Biochim. biophys. Acta* **226**, 103–12.

Floyd, R. A. & Soong, L. M. (1977). Obligatory free radical intermediate in the oxidative activation of the carcinogen N-hydroxy-2-acetylaminofluorene. *Biochim. biophys. Acta* **498**, 244–9.

Floyd, R. A., Soong, L. M., Stuart, M. A. & Reigh, D. C. (1978). Free radicals and carcinogenesis. Some properties of the nitroxyl free radicals produced by covalent binding of 2-nitrosofluorene to unsaturated lipids of membranes. *Arch. Biochem. Biophys.* **185**, 460–7.

Förster, T. (1946). Energiewanderung und Fluoreszenz. *Naturwissenschaften* **33**, 166–75.

Fowler, R. H. & Nordheim, L. (1928). Electron emission in intense electric fields. *Proc. R. Soc. Lond.* A **119**, 173–81.

Franck, J. (1925). Elementary processes of photochemical reactions. *Trans. Faraday Soc.* **21**, 536–42.

Frank, A. J., Grätzel, M., Henglein, A. & Janata, E. (1976). Electron transfer reactions of singlet and triplet pyrene in micelles with various radical anions in aqueous solution. *Ber. Bungsenges. phys. Chem.* **80**, 294–300.

Franz, W. (1967). Duration of the tunneling single process. *Phys. Status Solidi* **22**, K 139–40.

Frauenfelder, H. (1978). Principles of ligand binding to heme proteins. In *Methods in Enzymology* **54** E (ed. S. Fleischer and L. Packer), pp. 506–32. New York: Academic Press.

Frauenfelder, H. (1979). Molecular tunneling in heme proteins. **T**, pp. 627–46.

Frauenfelder, H., Petsko, G. A. & Tsernoglou, D. (1979). X-ray diffraction reveals protein structural dynamics. *Nature, Lond.* **280**, 558–63.

Freed, K. F. (1978). Radiationless transitions in molecules. *Acc. Chem. Res.* **11**, 74–80.

Freed, K. F. & Jortner, J. (1970). Multiphonon processes in the nonradiative decay of large molecules. *J. Chem. Phys.* **52**, 6272–91.

Frenkel, A. W. (1970). Bacterial photosynthesis. *Biol. Rev.* **45**, 569–616.

Frenkel, J. & Joffé, A. (1932). On the electric and photoelectric properties of contacts between a metal and a semiconductor. *Phys. Z. SowjU* **1**, 60–87.

Fröhlich, H. (1968). Long-range coherence and energy storage in biological systems. *Int. J. Quantum Chem.* **2**, 641–9.

Frumkin, A. N. (1933). Diskussionsbemerkungen. Following R. W. Gurney: Quantum mechanics and overpotential. *Phys. Z. SowjU* **4**, 360–4.

Fulton, T. A., Dynes, R. C. & Anderson, P. W. (1973). The flux shuttle – a Josephson junction shift register employing single flux quanta. *Proc. IEEE* **61**, 28–35.

Fuoss, R. M. (1958). Ionic association. III. The equilibrium between ion pairs and free ions. *J. Am. chem. Soc.* **80**, 5059–61.

Gamow, G. (1928). Zur Quanten theorie des Atom kernes. *Z. Phys.* **51**, 204–12.

Giaever, I. (1960*a*). Energy gap in superconductors measured by electron tunneling. *Phys. Rev. Lett.* **5**, 147–8.

Giaever, I. (1960*b*). Electron tunneling between two superconductors. *Phys. Rev. Lett.* **5**, 464–6.

Giaever, I. (1974). Electron tunneling and superconductivity. *Science, NY* **183**, 1253–8.

Giaever, I., Hart, H. R., Jr & Megerle, K. (1962). Tunneling into superconductors at temperatures below – 1 °K. *Phys. Rev.* **126**, 941–8.

Glaeser, R. M. & Berry, R. S. (1966). Mobilities of electrons and holes in organic molecular solids. Comparison of band and hopping models. *J. chem. Phys.* **44**, 3797–810.

Goldanskii, V. I. (1975). The low-temperature quantum-mechanical limit for the rate of chemical reactions. *Russ. Chem. Rev.* **44**, 1019–35 (trans. from *Usp. Khim.* **44**, 2121–49 (1975).)

Goldanskii, V. I. (1976). Chemical reactions at very low temperatures. *A. Rev. phys. Chem.* **27**, 85–126.

Goldanskii, V. I. (1979*a*). Quantum chemical reactivity near absolute zero: biological, chemical and astrophysical aspects. **T**, pp. 663–711.

Goldanskii, V. I. (1979*b*). Facts and hypotheses of molecular chemical tunneling. *Nature, Lond.* **279**, 109–15.

Goldanskii, V. I., Frank-Kamenetskii, M. D. & Barkalov, I. M. (1973). Quantum low-temperature limit of a chemical reaction rate. *Science, NY* **182**, 1344–5.

Gouterman, M. (1962). Radiationless transitions: a semiclassical model. *J. chem. Phys.* **36**, 2846–53.

Govindjee (ed.) (1975). *Bioenergetics of Photosynthesis.* New York: Academic Press.

Govindjee (ed.) (1982). *Photosynthesis. Vol I: Energy Conversion by Plants and Bacteria. Vol. II: Development, Carbon Metabolism, and Plant Productivity*, New York: Academic Press.

Govindjee, R., Becher, B. & Ebrey, T. G. (1978). The fluorescence from the chromatophore of the purple membrane protein. *Biophys. J.* **22**, 67–77.

Govindjee, R., Smith, W. R. & Govindjee (1974). Interaction of viologen dyes with chromatophores and reaction-center preparations from *Rhodospirillum rubrum. Photochem. & Photobiol.* **20**, 191–9.

Govindjee, R. & Sybesma, C. (1972). The photoreduction of nicotinamide-adenine dinucleotide by chromatophore fractions from *Rhodospirillum rubrum. Biophys. J.* **12**, 897–908.

Grätzel, M., Henglein, A. & Janata, E. (1975). Mechanism of electron transfer from e_{aq}^{-} to acceptors in micelles. *Ber. Bunsenges. phys. Chem.* **79**, 475–80.

Gregory, A. R., Henneker, W. H., Siebrand, W. & Zgierski, M. Z. (1976). Exactly solvable models for vibronic coupling in molecular spectroscopy. I. Nontotally symmetric harmonic mode. *J. chem. Phys.* **65**, 2071–87.

Grigorov, L. N. & Chernavskii, D. S. (1972). Quantum mechanical model of electron transfer from cytochrome to chlorophyll in photosynthesis. *Biofizika* **17**, 195–202 (202–9 in English translation).

Grigorov, L. N., Kononenko, A. A. & Rubin, A. B. (1970). Issledovanie nigkotemperaturnikh fotoindutsirovannikh okislitelnovosstanovitelnikh reaktsii tsitokhromov v kletkakh serobakterii *rhodopseudomonas* sp. *Mol. Biol. USSR* **4**, 483–90.

Gupta, R. K. & Yonetani, T. (1973). Nuclear magnetic resonance study of the interaction of cytochrome *c* with cytochrome *c* peroxidase. *Biochim. biophys. Acta* **292**, 502–8.

Gurney, R. W. (1931). The quantum mechanics of electrolysis. *Proc. R. Soc. Lond.* A **134**, 137–54.

Gurney, R. W. & Condon, E. U. (1928). Wave mechanics and radioactive disintegration. *Nature, Lond.* **122**, 439.

Gutmann, F. (1968). Electron tunneling in biological structures. *Nature, Lond.* **218**, 1359.

Hales, B. J. (1976). Temperature dependency of the rate of electron transport as a monitor of protein motion. *Biophys. J.* **16**, 471–80.

Halpern, E. H. & Wolf, A. A. (1972). Speculations of superconductivity in biological and organic systems. In *Advances in Cryogenic Engineering* (ed. K. D. Timmerhaus), **17**, 109–15. New York: Plenum Press.

Hansma, P. K. & Coleman, R. V. (1974). Spectroscopy of biological compounds with inelastic electric tunneling. *Science, NY* **184**, 1369–71.

Hart, E. J. & Anbar, M. (1970). *The Hydrated Electron.* New York: Wiley-Interscience.

Hayashi, T., Mataga, N., Umemoto, T., Sakata, Y. & Misumi, S. (1977*a*). Solvent-induced polarization phenomena in the excited state of composite systems with identical halves. 2. Effects of polarity upon the fluorescence of [2.2] (1,3) pyrenophane. *J. phys. Chem.* **81**, 424–9.

Hayashi, T., Suzuki, T., Mataga, N., Sakata, Y. & Misumi, S. (1977*b*). Solvent-induced polarization phenomena in the excited state of composite systems with identical halves. 1. Effects of solvent medium on the fluorescence spectra of 1,2-dianthrylethanes. *J. phys. Chem.* **81**, 420–3.

Heitler, W. & London, F. (1927). Wechselwirkung neutraler Atome und homöopolare Bindung nach der Quantenmechanik. *Z. Phys.* **44**, 455–72.

Henglein, A. (1974). Estimated distributions of electronic redox levels in aq/e^-_{aq} , H^+_{aq}/H_{aq} and some other systems. *Ber. Bunsenges. phys. Chem.* **78**, 1078–84.

Henglein, A. (1975). Estimated distribution functions of electronic redox levels and the rate of chemical reactions of excess electrons in dielectric liquids. *Ber. Bunsenges. Ges. phys. Chem.* **79**, 129–35.

Henry, B. R. & Siebrand, W. (1973). Radiationless transitions. In *Organic Molecular Photophysics*, vol. 1 (ed. J. B. Birks). New York: Wiley.

Herzberg, G. (1950). Molecular spectra and molecular structure. New York: Van Nostrand.

Herzberg, G. & Teller, E. (1933). Schwingungsstructur der Elektronenubergange bei mehratomigen Molekulen. *Z. phys. Chem.* (Leipzig) **21**, 410–46.

Hinatu, J., Masuhara, H., Mataga, N., Sakata, Y. & Misumi, S. (1978). Absorption spectra of inter- and intramolecular exciplex systems of pyrene and *N,N*-dimethylanaline in alcoholic solutions. *Bull. chem. Soc. Japan* **51**, 1032–6.

Hiyama, T. & Ke, B. (1971). A new photosynthetic pigment, 'P430': its possible role as the primary acceptor of photosystem. I. *Proc. natn. Acad. Sci. USA* **68**, 1010–13.

Hobey, W. D. & McLachlan, A. D. (1960). Dynamical Jahn–Teller effect in hydrocarbon radicals. *J. chem. Phys.* **33**, 1695–703.

Hodges, H. L., Holwerda, R. A. & Gray, H. B. (1974). Kinetic studies of the reduction of ferricytochrome *c* by $Fe(EDTA)^{2-}$. *J. Am. chem. Soc.* **96**, 3132–7.

Hoff, A. J. (1981). Magnetic field effects on photosynthetic reactions. *Quart. Rev. Biophys.* **14**, 599–665.

Hoff, A. J., Rademaker, H., Van Grondelle, R. & Duysens, L. N. M. (1977). On the magnetic field dependence of the yield of the triplet state in reaction centers of photosynthetic bacteria. *Biochim. biophys. Acta* **460**, 547–54.

Hoffmann, R. (1971). Interaction of orbitals through space and through bonds. *Acc. Chem. Res.* **4**, 1–9.

Hoffmann, R., Imamura, A. & Hehre, W. J. (1968). Benzynes, dehydroconjugated molecules, and the interaction of orbitals separated by a number of intervening σ bonds. *J. Am. chem. Soc.* **90**, 1499–509.

Holonyak, N, Jr, Lesk, I. A., Hall, R. N., Tiemann, J. J. & Ehrenreich, H. (1959). Direct observation of phonons during tunneling in narrow junction diodes. *Phys. Rev. Lett.* **3**, 167–8.

Holstein, T. (1952). Mobilities of positive ions in their parent gases. *J. Phys. Chem.* **56**, 832–6.

Holstein, T. (1959*a*). Studies of polaron motion. I. The molecular crystal model. *Ann. Phys.* (*NY*) **8**, 325–42.

Holstein, T. (1959*b*). Studies of polaron motion. II. The 'small' polaron. *Ann. Phys.* (*NY*) **8**, 343–89.

Holstein, T. (1979). Chemical-rate theory of small-polaron hopping. **T**, pp. 129–36.

Holten, D., Gouterman, M., Parson, W. W., Windsor, M. W. & Rockley, M. G. (1976). Electron transfer from photoexcited singlet and triplet bacteriopheophytin. *Photochem. & Photobiol.* **23**, 415–23.

Holten, D., Windsor, M. W., Parson, W. W. & Gouterman, M. (1978*a*). Models for bacterial photosynthesis: electron transfer from photoexcited singlet bacteriopheophytin to methyl viologen and *m*-dinitrobenzene. *Photochem. & Photobiol.* **28**, 951–61.

Holten, D., Windsor, M. W., Parson, W. W. & Thornber, J. P. (1978*b*). Primary photochemical processes in isolated reaction centres of *Rhodopseudomonas viridis*. *Biochim. biophys. Acta* **501**, 112–26.

Honig, B., Ebrey, T., Callender, R. H., Dinur, U. & Ottolenghi, M. (1979). Photoisomerization, energy storage, and charge separation: A model for light energy transduction in visual pigments and bacteriorhodopsin. *Proc. natn. Acad. Sci. USA* **76**, 2503–7.

Hopfield, J. J. (1974). Electron transfer between biological molecules by thermally activated tunneling. *Proc. natn. Acad. Sci. USA* **71**, 3640–4.

Hopfield, J. J. (1976). Fundamental aspects of electron transfer in biological membranes. In *Phénomènes èlectrique au niveau des membrane biologique. Proc. 29th Int. Congr. Société Chimie Physique* (ed. E. Roux), pp. 471–92. Amsterdam: Elsevier.

Hopfield, J. J. (1977). Photo-induced charge transfer. A critical test of the mechanism and range of biological electron transfer process. *Biophys. J.* **18**, 311–21.

Hopfield, J. J. (1979). Nonadiabatic electron tunneling: implications for bacterial photosynthesis and for critical tests of the mechanism. **T**, pp. 417–30.

Horiuti, J. & Polanyi, M. (1935). Grundlinien einer Theorie der Protonübertragung. Elektrolytische Dissoziation; Prototropie; spontane Ionisation und elektrolytische Abscheidung von Wasserstoff an Elektroden; Wasserstoffionenkatalyse. *Acta Electrochemica USSR* **2**, 505–32.

Hornbeck, J. A. (1952). Charge transfer and the mobility of rare gas ions. *J. phys. Chem.* **56**, 829–31.

Horne, R. A. (1963). Kinetics of the iron (II)–iron (III) electron exchange reaction in ice media. *J. inorg. nucl. Chem.* **25**, 1139–46.

Hsi, E. S. P. & Bolton, J. R. (1974). Flash photolysis-electron spin resonance study of the effect of o-phenanthroline and temperature on the decay time of the ESR signal B1 in reaction-center preparations and chromatophores of mutant and wild strains of *Rhodopseudomonas spheroides* and *Rhodospirillum rubrum. Biochim. biophys. Acta* **347**, 126–33.

Huang, K. & Rhys, A. (1950). Theory of light absorption and non-radiative transitions in F-centres. *Proc. R. Soc. Lond.* A **204**, 406–23.

Hui Bon Hoa, G., Begard, E., Debey, P. & Gunsalus, I. C. (1978). Two univalent electron transfers from putidaredoxin to bacterial cytochrome P-450 at subzero temperature. *Biochemistry* **17**, 2835–9.

Hund, F. (1927). Zur Deutung der Molekelspektern. III. Bemerkungen über das Schwingungs- und Rotationsspectrum bei Molekeln mit mehr als zwei Kernen. *Z. Phys.* **43**, 805–26.

Hunt, J. W. (1976). Early events in radiation chemistry. In *Advances in Radiation Chemistry* **5** (ed. M. Burton and J. L. Magee), pp. 185–315. New York: Wiley-Interscience.

Huppert, D., Rentzepis, P. M. & Tollin, G. (1976). Piocosecond kinetics of chlorophyll and chlorophyll/quinone solutions in ethanol. *Biochim. biophys. Acta* **440**, 356–64.

Hush, N. S. (1958). Adiabatic rate processes at electrodes. I. Energy–charge relationships. *J. chem. Phys.* **28**, 962–72.

Hush, N. S. (1961). Adiabatic theory of outer sphere electron-transfer reactions in solution. *Trans. Faraday Soc.* **57**, 577–80.

Hush, N. S. (1967). Intervalence-transfer absorption. 2. Theoretical considerations and spectroscopic data. *Prog. Inorg. Chem.* **8**, 391–444.

Isied, S. S. & Kuehn, C. (1979). Electron transfer properties of the imidazolate anion. **T**, pp. 229–36.

Isied, S. S. & Taube, H. (1973) Rates of intramolecular electron transfer. *J. Am. chem. Soc.* **95**, 8198–200.

Itoh, S. (1978). Membrane surface potential and the reactivity of the system II primary electron acceptor to charged electron carriers in the medium. *Biochim. biophys. Acta* **504**, 324–40.

Itoh, S. (1980a). Effects of surface potential and membrane potential on the midpoint potential of cytochrome *c*-555 bound to the chromatophore membrane of *Chromatium vinosum*. *Biochim. biophys. Acta* **591**, 346–55.

Itoh, S. (1980b). Study of electrogenic electron transfer steps in chromatophore membrane of *Chromatium vinosum* by the response of merocyanin dye. *Biochim. biophys. Acta* **593**, 212–23.

Jacks, C. A., Bennett, L. E., Raymond, W. N. & Lovenberg, W. (1974). Electron transfer to clostridial rubredoxin: kinetics of the reduction by hexaamineruthenium (II); vanadous and chromous ions. *Proc. natn. Acad. Sci. USA* **71**, 1118–22.

Jaklevic, R. C. & Lambe, J. (1966). Molecular vibration spectra by electron tunneling. *Phys. Rev. Lett.* **17**, 1139–40.

Jordan, P. (1927*a*). Über eine neue Begründung der Quantenmechanik. *Z. Phys.* **40**, 809–38.

Jordan, P. (1927*b*). Über eine neue Begrundung der Quantenmechanik: II. *Z. Phys.* **44**, 1–25.

Jortner, J. (1976). Temperature dependent activation energy for electron transfer between biological molecules. *J. chem. Phys.* **64**, 4860–7.

Jortner, J. (1980). Dynamics of the primary events in bacterial photosynthesis. *J. Am. chem. soc.* **102**, 6676–86.

Jortner, J., Rice, S. A. & Hochstrasser, R. M. (1969). Radiationless transitions in photochemistry. *Adv. Photochem.* **7**, 149–309.

Jortner, J. & Ulstrup, J. (1979). Dynamics of non-adiabatic atom transfer in biological systems. Carbon monoxide binding to hemoglobin. *J. Am. chem. Soc.* **101**, 7744–54.

Josephson, B. D. (1962). Possible new effects in superconductive tunneling. *Phys. Lett.* **1**, 251–3.

Josephson, B. D. (1974). The discovery of tunneling supercurrents. *Science, NY* **184**, 527–30.

Junge, W. & DeVault, D. (1975). Symmetry, orientation and rotational mobility in the a_3 heme of cytochrome *c* oxidase in the inner membrane of mitochondria. *Biochim. biophys. Acta* **408**, 200–14.

Jursinic, P. & Govindjee (1977). Temperature dependence of delayed light emission in the 6 to 340 microsecond range after a single flash in chloroplasts. *Photochem. & Photobiol.* **26**, 617–28.

Kakitani, T. & Kakitani, H. (1981). A possible new mechanism of temperature dependence of electron transfer in photosynthetic system. *Biochim. biophys. Acta* **635**, 498–514.

Kampars, V. & Neilands, O. (1977). The electron affinities of organic electron acceptors. *Russ. chem. Rev.* **46**, 503–13. (*Usp. Khim.* **46**, 945–66.)

Kano, K., Takuma, K., Ikeda, T., Nakajima, D., Tsutsui, Y. & Matsuo, T. (1978). Zinc tetraphenylporphyrin-sensitized photoreduction of anthraquinonesulfonate in aqueous micellar solutions. *Photochem. & Photobiol.* **27**, 695–701.

Kaufmann, K. J., Dutton, P. L., Netzel, T. L., Leigh, J. S. & Rentzepis, P. M. (1975).

Picosecond kinetics of events leading to reaction center bacteriochlorophyll oxidation. *Science, NY* **188**, 1301–4.

Kaufmann, K. J., Petty, K. M., Dutton, P. L. & Rentzepis, P. M. (1976). Picosecond kinetics in reaction centres of *Rps. sphaeroides* and the effects of ubiquinone extraction and reconstitution. *Biochem. biophys. Res. Commun.* **70**, 839–45.

Ke, B., Demeter, S., Zamaraev, K. I. & Khairutdinov, R. F. (1979*a*). Electron tunneling in Photosystem-I charge recombination at low temperatures. **T**, pp. 371–6.

Ke, B., Demeter, S., Zamaraev, K. I. & Khairutdinov, R. F. (1979*b*). Charge recombination in photosystem I at low temperatures: kinetics of electron tunneling. *Biochim. biophys. Acta* **545**, 265–84.

Ke, B., Hansen, R. E. & Beinert, H. (1973). Oxidation-reduction potentials of bound iron-sulfur proteins of photosystem I. *Proc. natn. Acad. Sci. USA* **70**, 2941–5.

Kennel, S. J. & Kamen, M. D. (1971). Iron-containing proteins in *Chromatium*. II. Purification and properties of cholate-solubilized cytochrome complex. *Biochem. biophys. Acta* **253**, 153–66.

Kestner, N. R., Logan, J. & Jortner, J. (1974). Thermal electron transfer reactions in polar solvents. *J. Phys. Chem.* **78**, 2148–66.

Kevan, K. (1974). Trapped electrons in organic glasses. In *Advances in Radiation Chemistry* **4** (ed. M. Burton and J. L. Magee), pp. 181–305. New York: Wiley-Interscience.

Kihara, T. & Chance, B. (1969). Cytochrome photooxidation at liquid nitrogen temperatures in photosynthetic bacteria. *Biochim. biophys. Acta* **189**, 116–24.

Kihara, T., DeVault, D. & Chance, B. (1969). Light-induced reactions at liquid nitrogen temperatures in photosynthetic bacteria. *13th Biophys. Soc. Mtg.* (Los Angeles), Abstr. WPM-D5.

Kihara, T. & Dutton, P. L. (1970). Light-induced reactions of photosynthetic bacteria. I. Reactions in whole cells and in cell-free extracts at liquid nitrogen temperatures. *Biochem. biophys. Acta* **205**, 196–204.

Kihara, T. & McCray, J. A. (1973). Water and cytochrome oxidation-reduction reactions. *Biochim. biophys. Acta* **292**, 297–309.

Kiryukhin, D. P., Barkalov, I. M. & Goldanskii, V. I. (1979). Low temperature limit of the rate of propagation of chains in the hydrobromination of ethylene. *J. chim. phys.* **76**, 1013–15.

Kitagawa, T., Ozaki, Y. & Kyogoku, Y. (1978). Resonance Raman studies on the ligand–iron interactions in hemoproteins and metallo–porphyrins. *Adv. Biophys.* **11**, 153–96.

Klein, J., Léger, A., Belin, M., Defourneau, D. & Sangster, M. J. L. (1973). Inelastic-electron-tunneling spectroscopy of metal-insulator-metal junctions. *Phys. Rev.* B **7**, 2336–48.

Klingenberg, M. & Schollmeyer, P. (1963). Redox reactions in mitochondria under the control of ATP. *Proc. 5th Int. Congr. Biochem.*, Moscow **5**, pp. 46–65. London: Pergamon.

Knaff, D. B. & Arnon, D. I. (1969). Light-induced oxidation of a chloroplast b-type cytochrome at − 189 °C. *Proc. natn. Acad. Sci. USA* **63**, 956–62.

Knaff, D. B. & Malkin, R. (1974). The effect of temperature on the primary reaction of chloroplast photosystem II. Evidence for a temperature-dependent back reaction. *Biochim. biophys. Acta* **347**, 395–403.

Knox, R. S. (1979). Localized and delocalized tunneling states. **T**, pp. 91–4.

Koido, S. (1960). Uber die Berechung von Franck–Condon–Integralen. *Z. Naturforsch.* **15a**, 123–8.

Kononenko, A. A., Lukashev, E. P., Rubin, A. B. & Venediktov, P. S. (1972). On the interaction of photoactive bacteriochlorophyll with the primary electron acceptor in the reaction centre of *Ectothiorhodospira shaposhnikovii. Biochim. biophys. Acta* **275**, 130–3.

Koppenol, W. H., VanBuren, K. J. H., Butler, J. & Braams, R. (1976). The kinetics of the reduction of cytochrome *c* by the superoxide anion radical. *Biochem. biophys. Acta* **449**, 457–68.

Korman, C. S. & Coleman, R. V. (1977). Inelastic electron tunneling spectroscopy of single ring compounds adsorbed on alumina. *Phys. Rev.* B **15**, 1877–93.

Kowalsky, A. (1965). Nuclear magnetic resonance studies of cytochrome *c*. Possible electron delocalization. *Biochemistry* **4**, 2382–8.

Kramers, H. A. (1926). Wellenmechanik und halbzahlige Quantisierung. *Z. Phys.* **39**, 828–40.

Kramers, H. A. (1934). L'interaction entre les atomes magnétogènes dans un cristal paramagnétique. *Physica, 's Grav.* **1**, 182–92.

Krebs, H. & Johnson, W. A. (1937). The role of citric acid in intermediate metabolism in animal tissues. *Enzymologia* **4**, 148–56.

Kubo, R. (1952). Thermal ionization of trapped electrons. *Phys. Rev.* **86**, 929–37.

Kubo, R. & Toyozawa, Y. (1955). Application of the method of generating function to radiative and non-radiative transitions of a trapped electron in a crystal. *Prog. theor. Phys.* **13**, 160–82.

Kuhn, H., Möbius, D. & Bücher, H. (1972). Spectroscopy of monolayer assemblies. I: Principles and applications. II: Experimental procedure. In *Physical Methods of Chemistry*, part IIIB (ed. A. Weissberger and B. W. Rossiter), pp. 577–702. New York: Wiley-Interscience.

Kung, M. C. & DeVault, D. (1976). Carotenoid triplet state in *R. Spheroides* GA chromatophores. *Photochem. & Photobiol.* **24**, 87–9.

Kurtin, S. L., McGill, T. C. & Mead, C. A. (1971). Direct inter-electrode tunneling in GaSe. *Phys. Rev.* B **3**, 3368–79.

Kuznetsov, A. M., Søndergård, N. C. & Ulstrup, J. (1978). Low-temperature electron transfer in bacterial photosynthesis. *Chem. Phys.* **29**, 383–90.

Ladik, J. (1974). The energy band structure and conduction properties of DNA. *Int. J. Quantum Chem., Quantum Biol. Symp.* **1**, 65–9.

Ladik, J. (1975). Toward the electronic structure of real DNA. *Int. J. Quantum Chem., Quantum Biol. Symp.* **2**, 133–43.

Ladik, J. (1978). Electronic structure of biopolymers, pp. 257–78; Energy band structure of highly conducting polymers. pp. 279–88. In *Quantum Theory of Polymers* (ed. J. M. Andre, J. Delhalle and J. Ladik). Dordrecht: D. Reidel.

Ladik, J., Biczo, G. & Redly, J. (1969). Possibility of superconductor-type enhanced conductivity in DNA at room temperature. *Phys. Rev.* **188**, 710–15.

Laki, K. & Ladik, J. (1976). A note on the 'electronic theory' of cancer. In *Int. J. Quantum Chem., Quantum Biol. Symp.* **3**, 51–7.

Land, E. J. & Swallow, A. J. (1971). One-electron reactions in biochemical systems as studied by pulse radiolysis. V. Cytochrome C. *Arch. Biochem. Biophys.* **145**, 365–72.

Landau, L. (1932). Zur Theorie der Energieübertragung bei Stössen. *Phys. Z. SowjU* **1**, 88–98.

Laplante, J.-P. & Siebrand, W. (1978). Non-Arrhenius behaviour of transitions across energy barriers due to crossing of potential energy surfaces. *Chem. Phys. Lett.* **59**, 433–6.

Lax, M. (1952). The Franck–Condon principle and its application to crystals. *J. chem. Phys.* **20**, 1752–60.

Lee, C.-Y. (1980). The dipole flip-flop theory for the excitation and conduction of excitable tissues. I. Dipole's behavior and gating currents. *Bull. Math. Biol.* **42**, 489–505.

Leigh, J. S., Jr & Wilson, D. F. (1972). Heme–heme interactions in cytochrome *c* oxidase; effects of photodissociation of the CO compound. *Biochem. biophys. Res. Commun.* **48**, 1266–72.

Leigh, J. S., Jr, Wilson, D. F., Owen, C. S. & King, T. E. (1974). Heme–heme interaction in cytochrome *c* oxidase: the cooperativity of the hemes of cytochrome *c* oxidase as evidenced in the reaction with CO. *Archs Biochem. Biophys.* **160**, 476–86.

Levich, V. G. (1966). Present state of the theory of oxidation-reduction in solution (bulk and electrode reactions). *Adv. Electrochem. & Electrochem. Eng.* **4**, 249–371.

Levich, V. G. (1970). Kinetics of reactions with charge transport. In *Physical Chemistry an Advanced Treatise*. Vol. IX. *B/Electrochemistry* (ed. H. Eyring), pp. 986–1074.

Levich, V. G. & Dogonadze, R. R. (1959). Teoriya bezizluchatelnikh electronnikh perekhodov mezhdu ionami v rastvorakh. *Dokl. Acad. Nauk. SSSR* **124**, 123–6.

Levich, V. G. & Dogonadze, R. R. (1961). Adiabaticheskaya teoriya elektronnikh protsessov v rastorakh. *Coll. Czech. chem. Commun.* **26**, 193–214.

Lewis, T. J. (1977). The dielectric behaviour of non-crystalline solids. In *Dielectric and Related Molecular Processes*, Vol. 3, pp. 186–218 (ed. M. Davies). London: The Chemical Society.

Libby, W. F. (1940). Reactions of high energy atoms produced by slow neutron capture. *J. Am. chem. Soc.* **62**, 1930–43.

Libby, W. F. (1952). Theory of electron exchange reactions in aqueous solution. *J. Phys. Chem.* **56**, 63–8.

Libby, W. F. (1979). My tunneling experiences. **T**, pp. 3–8.

Likhtenstein, G. I., Kulikov, A. V., Kotelnikov, A. I. & Bogatyrenko, V. R. (1982). Structure and action mechanism of reaction centers of photosynthetic bacteria. *Photobiochem. Photobiophys.* **3**, 337–44.

Lin, L. & Thornber, J. P. (1975). Isolation and partial characterization of the photochemical reaction center of *Chromatium vinosum* (Strain D). *Photochem. & Photobiol.* **22**, 37–40.

Lin, S. H. (1966). Rate of interconversion of electronic and vibrational energy. *J. chem. Phys.* **44**, 3759–67.

Lin, S. H. & Bersohn, R. (1968). Effect of partial deuteration and temperature on triplet-state lifetimes. *J. chem. Phys.* **48**, 2732–6.

Lindquist, N. G. (1973). *ACIA Radiol. Supplement 325*, pp. 1–92.

Little, W. A. (1964). Possibility of synthesizing an organic superconductor. *Phys. Rev.* **134A**, 1416–24.

Loach, P. A., Kung, M. C. & Hales, B. J. (1975). Characterization of the phototrap in photosynthetic bacteria. *Ann. NY Acad. Sci.* **244**, 297–319.

London, F. J. (1937). Quantum theory of interatomic currents in aromatic compounds. *J. Phys. Radium* **8**, 397–409.

Löwdin, P-O. (1963). Proton tunneling in DNA and its biological implications. *Rev. mod. Physics* **35**, 724–32.

Lozier, R. H. & Butler, W. L. (1974). Light-induced absorbance changes in chloroplasts mediated by photosystem I and photosystem II at low temperature. *Biochim. biophys. Acta* **333**, 465–80.

Lukashev, E. P., Timofeev, K. N., Kononenko, A. A., Venediktov, P. S. & Rubin, A. B. (1976). Temperature dependence of the reduction kinetics of photooxidized reaction center bacteriochlorophyll in dark and light adapted chromatophores of purple bacteria. *Photosynthetica* **10**, 423–30.

McConnell, H. M. (1961). Intramolecular charge transfer in aromatic free radicals. *J. chem. Phys.* **35**, 508–15.

McCray, J. A. & Kihara, T. (1979). Rates of reduced cytochrome *c*–ferricyanide binding and electron transfer. *Biochim. biophys. Acta* **548**, 417–26.

McElroy, J. D., Feher, G. & Mauzerall, D. C. (1969). On the nature of the free radical formed during the primary process of bacterial photosynthesis. *Biochim. biophys. Acta* **172**, 180–3.

McElroy, J. D., Mauzerall, D. C. & Feher, G. (1974). Characterization of primary reactants in bacterial photosynthesis. II. Kinetic studies of the light-induced EPR signal (g = 2.0026) and the optical absorbance changes at cryogenic temperatures. *Biochim. biophys. Acta* **333**, 261–77.

McGinness, J. & Proctor, P. (1973). The importance of the fact that melanin is black. *J. theor. Biol.* **39**, 677–8.

McGinness, J. E., Corry, P. M. & Proctor, P. H. (1974). Amorphous semiconductor switching in melanins. *Science* **183**, 853–5.

Malkin, R. & Bearden, A. J. (1971). Primary reactions of photosynthesis: photoreduction of a bound chloroplast ferredoxin at low temperature as detected by ESR spectroscopy. *Proc. natn. Acad. Sci. USA* **68**, 16–19.

Mann, B. & Kuhn, H. (1971). Tunneling through fatty acid salt monolayers. *J. appl. Phys.* **42**, 4398–405.

Manneback, C. (1951). Computation of the intensities of vibrational spectra of electronic bands in diatomic molecules. *Physica, 's Grav* **17**, 1001–10.

Marcolin, H. E., Reschke, R. & Trautwein, A. (1977). Untersuchung der Recombinations kinetik von photodissoziiertem Myoglobin-CO bei tiefen Temperaturen mittels Mössbauerspektroskopie, *Z. Naturforsch.* **32 C**, 683–95.

Marcus, R. A. (1956). On the theory of oxidation-reduction reactions involving electron transfer: I. *J. chem. Phys.* **24**, 966–78.

Marcus, R. A. (1965). On the theory of electron-transfer reactions. VI. Unified treatment for homogeneous and electrode reactions. *J. chem. Phys.* **43**, 679–701.

Marcus, R. A. (1968). Electron transfer reactions. In *Chemische Elementarprozesse* (ed. H. Hartmann), pp. 348–56. Berlin: Springer-Verlag.

Marcus, R. A. (1979*a*). Electron and nuclear tunneling in chemical and biological systems. **T**, pp. 109–27.

Marcus, R. A. (1979*b*). Electron transfer and tunneling in chemical and biological systems. In *Light-induced Charge Separation in Biology and Chemistry* (ed. H. Gerischer and J. J. Katz), pp. 15–43. Berlin: Dahlem Konferenzen, Verlag Chemie.

Marcus, R. A. & Siders, P. (1982). Theory of highly exothermic electron transfer reactions. *J. Phys. Chem.* **86**, 622–30.

Marcus, R. J., Zwolinski, B. J. & Eyring, H. (1954). The electron tunneling. hypothesis for electron exchange reactions. *J. Phys. Chem.* **58**, 432–7.

Mason, R. (1959). Charge transfer processes in biological systems. *Disc. Faraday Soc.* **27**, 129–33.

Mataga, N., Migita, M. & Nishimura, T. (1978). Picosecond chemistry of some exciplex systems. *J. Mol. Struct.* **47**, 199–219.

Mathews, J. & Walker, R. L. (1965). *Mathematical Methods of Physics*. New York: Benjamin.

Mathis, P. (1969). Triplet-triplet energy transfer from chlorophyll-*a* to carotenoids in solution and in chloroplasts. In *Progress in Photosynthesis Research*, vol. 2 (ed. H. Metzner), pp. 818–22. Tübingen.

Mathis, P. & Conjeaud, H. (1979). Rapid reduction of P-700 photooxidized by a flash at low temperature in spinach chloroplasts. *Photochem. & Photobiol.* **29**, 833–7.

Mathis, P., Haveman, P. & Yates, M. (1976). The reaction center of photosystem II. In *Brookhaven Symp. Biol.* **28** (ed. J. M. Olson and G. Hind), pp. 267–77.

Mathis, P., Sauer, K. & Remy, R. (1978). Rapidly reversible flash-induced electron transfer in a P-700 chlorophyll-protein complex isolated with SDS. *FEBS Lett.* **88**, 275–8.

Matsuura, K., Masamoto, K., Itoh, S. & Nishimura, M. (1979). Effect of surface potential on the intramembrane electrical field measured with carotenoid spectral shift in chromatophores from *Rhodopseudomonas sphaeroides*. *Biochim. biophys. Acta* **547**, 91–102.

Matsuura, K., Takamiya, K., Itoh, S. & Nishimura, M. (1980). Effects of surface potential on the equilibrium and kinetics of redox reactions of membrane components with external reagents in chromatophores from *Rhodoseudomonas sphaeroides*. *J. Biochem.* **87**, 1431–7.

Mauk, A. G., Scott, R. A. & Gray, H. B. (1980). Electron transfer distances to and from metalloprotein redox sites in reactions with inorganic complexes. *J. Am. chem. Soc.* **102**, 4360–3.

Mead, C. A. (1969). Some properties of exponentially damped wave functions. In *Tunneling phenomena in solids* (ed. E. Burstein and S. Lundqvist), pp. 127–34. New York: Plenum.

Merzbacher, E. (1961). *Quantum Mechanics*. New York: Wiley.

Migita, M., Kawai, M. & Mataga, N. (1978). Picosecond laser photolysis studies of intramolecular charge transfer and heteroexcimer formation processes. *Chem. Phys. Lett.* **53**, 67–70.

Miller, J. R. (1972). Scavenging kinetics of electrons produced by irradiation of organic glasses; experimental evidence for long range tunneling. *J. chem. Phys.* **56**, 5173–83.

Miller, J. R. (1973). Fast electron transfer reactions in a rigid matrix: further evidence for quantum mechanical tunneling. *Chem. Phys. Lett.* **22**, 180–2.

Miller, J. R. (1975). Reactions of trapped electrons by quantum mechanical tunneling observed by pulse radiolysis of an aqueous glass. *J. Phys. Chem.* **79**, 1070–8.

Miller, J. R. (1978). Tunneling reactions of trapped electrons with added electron acceptors in alcohol glasses at 77 °K. *J. Phys. Chem.* **82**, 767–74.

Miller, J. R., Beitz, J. V. & Huddleston, R. K. (1983). Effect of free energy on rates of electron transfer between molecules (submitted for publication).

Miller, J. R., Hartman, K. W. & Abrash, S. (1982). Long distance (25 Å) electron transfer by triplet excited states in rigid media. *J. Am. chem. Soc.* **104**, 4296–8.

Miller, J. R. & Willard, J. E. (1972). Time-dependent scavenging of trapped electrons in 2-methyl-tetrahydrofuran glass at 77 °K. *J. Phys. Chem.* **76**, 2641–2.

Monger, T. G., Cogdell, R. J. & Parson, W. W. (1976). Triplet states of the bacteriochlorophyll and carotenoids in chromatophores of photosynthetic bacteria. *Biochim. biophys. Acta* **449**, 136–53.

Morrison, L. E. & Loach, P. A. (1978). Complex charge recombination kinetics of the photo-trap in *Rhodospirillum rubrum*. *Photochem. & Photobiol.* **27**, 751–7.

Moscowitz, E. & Malley, M. M. (1978). Energy transfer and photooxidation kinetics in reaction centers on the picosecond time scale. *Photochem. & Photobiol.* **27**, 55–9.

Mott, N. F. (1939). A theory of the formation of protective oxide films on metals. *Trans. Faraday Soc.* **35**, 1175–7.

Mott, N. F. (1940). The theory of the formation of protective oxide films on metals: II. *Trans. Faraday Soc.* **36**, 472–83.

Mueller, P. & Rudin, D. O. (1963). Induced excitability in reconstituted cell membrane structure. *J. theor. Biol.* **4**, 268–80.

Müller, A. & Witt, H. T. (1961). Trapped primary product of photosynthesis in green plants. *Nature, Lond.* **189**, 944–5.

Mulliken, R. S. (1952). Molecular compounds and their spectra: II. *J. Am. chem. Soc.* **74**, 811–24.

Nakashima, N., Murakawa, M. & Mataga, N. (1976). Picosecond flash spectroscopy of solvent-induced intramolecular electron transfer in the excited 9,9′-bianthryl. *Bull. chem. Soc. Japan* **49**, 854–8.

Neiss, M. A., Sprague, E. D. & Willard, J. E. (1975). Decay mechanisms of CH_3 in 3-methylpentane and 3MP-d_{14} glasses at 77 °K as indicated by methane yields. *J. chem. Phys.* **63**, 1118–21.

Nicol, J., Shapiro, S. & Smith, P. H. (1960). Direct measurement of the superconducting energy gap. *Phys. Rev. Lett.* **5**, 461–4.

Nikitin, E. (1968). Theory of non-adiabatic transitions. Recent development of the Landau-Zener (linear) model. In *Chemische Elementarprozesse* (ed. H. Hartmann), pp. 43–77. New York: Springer-Verlag.

Nikitin, E. E. (1974). *Theory of Elementary Atomic and Molecular Processes in Gases* (trans. M. J. Kearsley), p. 115. Oxford: Clarendon.

Noks, P. P., Lukashev, E. P., Kononenko, A. A., Venediktov, P. S. & Rubin, A. B. (1977). O vozmoskonoi roli macromolekulyarnikh komponentov v funktsionirovannii fotosinteticheskikh reaktsionnikh tsentrov purpurnikh bakterii. *Mol. Biol. USSR* **11**, 1090–9.

Nordheim, L. (1928). Zur Theorie der thermischen Emission und der Reflexion von Electronen an Metallen. *Z. Phys.* **46**, 833–55.

Nordheim, L. (1932). Zur Theorie der Detector wirkung. *Z. Phys.* **75**, 434–41.

Norris, J. R., Bowman, M. K., Budil, D. E., Tang, J. Wraight, C. A. & Closs, G. L. (1982). Magnetic characterization of the primary state of bacterial photosynthesis. *Proc. natn. Acad. Sci. USA* **74**, 5532–6.

Norton, K. A., Hulett, L. G., Halko, D. J. & Hurst, J. K. (1979). Electron transfer in $(NH_3)_5$Ru(III)-4-alkylpyridine-Cu(I) binuclear ions. T, pp. 237–41.

Ohnishi, T. (1979). Mitochondrial iron sulfur flavo-dehydrogenases. In *Membrane Proteins in Energy Transduction* (ed. R. A. Capaldi), pp. 1–87. New York: Marcell-Decker.

Okada, T., Saito, T., Mataga, N., Sakata, Y. & Misumi, S. (1977). Solvent effects upon dynamic behavior of intramolecular hetero-excimers. *Bull. chem. Soc. Japan* **50**, 331–6.

Okamura, M. Y., Fredkin, D. R., Isaacson, R. A. & Feher, G. (1979*a*). Magnetic interactions and electron transfer kinetics of the reduced intermediate acceptor in reaction centers (RC's) of *Rhodopseudomonas sphaeroides* R-26. Evidence for thermally induced tunneling. **T**, pp. 729–43.

Okamura, M. Y., Isaacson, R. A. & Feher, G. (1979*b*). Spectroscopic and kinetic properties of the transient intermediate acceptor in the reaction centres of *Rhodopseudomonas sphaeroides*. *Biochim. biophys. Acta* **546**, 394–417.

Olson, J. M. & Chance, B. (1960). Oxidation-reduction reactions in the photosynthetic bacterium *Chromatium*. I. Absorption spectrum changes in whole cells. *Archs Biochem. Biophys.* **88**, 26–39.

Oppenheimer, J. R. (1928). Three notes on the quantum theory of aperiodic effects. *Phys. Rev.* **13**, 66–81.

Otto, P., Ladik, J., Laki, K. & Szent–Gyorgyi, A. (1978). Internal charge transfer in proteins to the Schiff bases of their lysine side chains. *Proc. natn. Acad. Sci. USA* **75**, 3548–50.

Otto, P., Suhai, S. & Ladik, J. (1977). *Ab initio* supermolecule study of charge transfer in the glyoxal-formamide and in the H_2S-formamide systems. *Int. J. Quantum Chem., Quantum Biol. Symp.* **4**, 451–7.

Padden–Row, M. N., Patney, H. K., Brown, R. S. & Houk, K. N. (1981). Observation of a very large orbital interaction through four bonds. An alternative model of orbital interactions through bonds. *J. Am. chem. Soc.* **103**, 5575–7.

Parak, F., Frolov, E. N., Kononenko, A. A., Mössbauer, R. L., Goldanskii, V. I. & Rubin, A. B. (1980). Evidence for a correlation between the photoinduced electron transfer and dynamic properties of the chromatophore membranes from *Rhodospirillum rubrum*. *FEBS Lett.* **117**, 368–72.

Parson, W. W. (1967). Flash-induced absorbance changes in *Rhodospirillum rubrum* chromatophores, *Biochim. biophys. Acta* **131**, 154–72.

Parson, W. W. (1968). The role of P870 in bacterial photosynthesis. *Biochim. biophys. Acta* **153**, 248–59.

Parson, W. W. (1969*a*). The reaction between primary and secondary electron acceptors in bacterial photosynthesis. *Biochim. biophys. Acta* **189**, 384–96.

Parson, W. W. (1969*b*). Cytochrome photooxidations in *Chromatium* chromatophores. Each P870 oxidizes two cytochrome C442 hemes. *Biochim. biophys. Acta* **189**, 397–403.

Parson, W. W. & Case, G. D. (1970). In *Chromatium*, a single photochemical reaction center oxidizes both cytochrome C552 and cytochrome C555. *Biochim. biophys. Acta* **205**, 232–45.

Parson, W. W., Clayton, R. K. & Cogdell, R. J. (1975). Excited states of photosynthetic reaction centers at low redox potentials. *Biochim. biophys. Acta* **387**, 265–78.

Pasman, P., Rob, F. & Verhoeven, J. W. (1982). Intramolecular charge-transfer absorption and emission resulting from through-bond interaction in bichromophoric molecules. *J. Am. chem. Soc.* **104**, 5127–33.

Pauling, L. (1928). The application of the quantum mechanics to the structure of the hydrogen molecule and hydrogen molecule ion and to related problems. *Chem. Rev.* **5**, 173–213.

Pauling, L. (1930). The rotational motion of molecules in crystals. *Phys. Rev.* **36**, 430–43.

Pauling, L. & Wilson, E. B. (1935). *Introduction to Quantum Mechanics*. New York: McGraw-Hill.

Pekar, S. I. (1951). *Investigations on the Electronic Theory of Crystals*. Moscow: Fizmatgiz. (In Russian.)

Pekar, S. I. (1954). *Untersuchungen Über die Electronen-theorie der Kristalle*. Berlin: Akademie-Verlag.

Peters, K., Avouris, P. & Rentzepis, P. M. (1978). Picosecond dynamics of primary electron-transfer process in bacterial photosynthesis. *Biophys. J.* **23**, 207–17.

Pethig, R. (1978). Electronic properties of protein-methylglyoxal complexes: strong evidence for energy band conduction. *Int. J. Quantum Chem., Quantum Biol. Symp.* **5**, 159–71.

Pethig, R. (1979). *Dielectric and Electronic Properties of Biological Materials.* New York: Wiley.

Pethig, R. & Szent–Györgyi, A. (1977). Electronic properties of the caseinmethylglyoxal complex. *Proc. natn. Acad. Sci. USA* **74**, 226–8.

Petrov, E. G. (1979). Mechanisms of electron transfer through proteins *Int. J. Quantum Chem.* **16**, 133–52.

Petrov, E. G., Teslenko, V. I. & Kharkyanen, V. N. (1978). The nature of the gating current carriers of sodium channels. An electronic model. *Dokl. Akad. SSSR* **238**, 967–70.

Potasek, M. J. (1979). Dynamics of electron transport in macromolecules. **T**, pp. 433–8.

Potasek, M. J. & Hopfield, J. J. (1977*a*). An experimental test of the vibronically coupled tunneling description of biological electron transfer. *Proc. natn. Acad. Sci. USA* **74**, 229–33.

Potasek, M. J. & Hopfield, J. J. (1977*b*). Fundamental aspects of electron transfer: experimental verification of vibronically coupled electron tunneling. *Proc. natn. Acad. Sci. USA* **74**, 3817–20.

Prince, R. C., Leigh, J. S., Jr & Dutton, P. L. (1976). Thermodynamic properties of the reaction center of *Rhodopseudomonas viridis.* In vivo measurement of the reaction center bacteriochlorophyll-primary acceptor intermediary electron carrier. *Biochim. biophys. Acta* **440**, 622–36.

Proceedings of the international conference in fluids. (September 1976) (1977). *Can. J. Chem.* **55**, 1795–2277.

Proctor, P. H. (1972). Electron transfer factors in psychosis and dyskinesia. *Physiol. Chem. & Phys.* **4**, 349–60.

Proctor, P. McGinness, J. & Corry, P. (1974). A hypothesis on the preferential destruction of melanized tissues. *J. theor. Biol.* **48**, 19–22.

Pullman, B. (1979). Recent developments on the mechanism of chemical carcinogenesis by aromatic hydrocarbons. *Int. J. Quantum Chem.* **16**, 669–89.

Pullman, B. & Pullman, A. (1958). Electron-donor and acceptor properties of biologically important purines, pyrimidines, pteridines, flavins and aromatic amino acids. *Proc. natn. Acad. Sci. USA* **44**, 1197–202.

Pullman, B. & Pullman, A. (1963). *Quantum Biochemistry.* New York: Wiley-Interscience.

Pyt'eva, N. F., Andreenko, T. I. & Rubin, A. A. (1976). Existence of cyclic and non-cyclic electron transport systems in bacterial photosynthetic apparatus. 2. Experimental evidence for the scheme of electron-transport reactions in the chromatophores of photosynthesizing bacteria *Ectothiorhodospira shaposhnikovii. Photosynthetica* **10** (1), 14–19.

Pyt'eva, N. F., Ratyni, A. I. & Rubin, A. B. (1976). Existence of cyclic and non-cyclic electron transport systems in bacterial photosynthetic apparatus. 1. Theoretical differences in kinetic characteristics of cyclic and non-cyclic systems of electron transport. *Photosynthetica* **10** (1), 7–13.

Randles, J. E. B. (1952). Kinetics of rapid electrode reactions. 2. Rate constants and activation energies of electrode reactions. *Trans. Faraday Soc.* **48**, 828–32.

Redi, M. & Hopfield, J. J. (1980). Theory of thermal and photoassisted electron tunneling. *J. chem. Phys.* **72**, 6651–60.

Reed, D. W.,& Clayton, R. K. (1968). Isolation of a reaction center fraction from *Rhodopseudomonas spheroides*. *Biochem. biophys. Res. Commun.* **30**, 471–5.

Rehm, D. & Weller, A. (1970). Kinetics of fluorescence quenching by electron and H-atom transfer. *Israel J. Chem.* **8**, 259–71.

Rice, O. K. (1967). *Statistical Mechanics, Thermodynamics and Kinetics*. San Francisco: Freeman.

Rice, S. A. & Pilling, M. J. (1978). Tunneling reactions of solvated electrons in liquids and glasses. *Prog. React. Kinet.* **9**, 93–194.

Rieder, K. & Taube, H. (1977). Intramolecular electron transfer. 3. Bis(4-pyridyl) methane as a bridging group. *J. Am. chem. Soc.* **99**, 7891–4.

Robin, M. B. & Day, P. (1967). Mixed valence chemistry – a survey and classification. *Adv. inorg. Chem. Radiochem.* **10**, 247–422.

Robinson, G. W. & Frosch, R. P. (1962). Theory of electronic energy relaxation in the solid phase. *J. Chem. Phys.* **37**, 1962–73.

Robinson, G. W. & Frosch, R. P. (1963). Electronic excitation transfer and relaxation. *J. chem. Phys.* **38**, 1187–203.

Rockley, M. G., Windsor, M. W., Cogdell, R. J. & Parson, W. W. (1975). Picosecond detection of an intermediate in the photochemical reaction of bacterial photosynthesis. *Proc. natn. Acad. Sci. USA* **72**, 2251–5.

Romijn, J. C. & Amesz, J. (1976). Photochemical activites of reaction centres from *Rhodopseudomonas sphaeroides* at low temperature and in the presence of chaotropic agents. *Biochim. biophys. Acta* **423**, 164–73.

Rosenberg, B. (1965). Some problems in electrical conductivity of proteins. In *Oxidases and Related Redox Systems* (ed. T. King, H. Mason and M. Morrison), pp. 72–87. New York: Wiley.

Rosenfeld, T., Honig, B., Ottolenghi, M., Hurley, J. & Ebrey, T. G. (1977). Cis–trans isomerization in the photochemistry of vision. *Pure & appl. Chem.* **49**, 431–51.

Rowell, J. M. (1963). Magnetic field dependence of the Josephson tunnel current. *Phys. Rev. Lett.* **11**, 200–2.

Rowell, J. M., Anderson, P. W. & Thomas, D. E. (1963). Image of the phonon spectrum in the tunneling characteristic between superconductors. *Phys. Rev. Lett.* **10**, 334–6.

Rowell, J. M., Chynoweth, A. G. & Phillips, J. C. (1962). Multiphonon effects in tunneling between metals and superconductors. *Phys. Rev. Lett.* **9**, 59–61.

Rubin, A. B., Kononenko, A. A., Venediktov, P. S., Borisevitch, G. P., Knox, P. P. & Lukashev, E. P. (1980). Polarization effects in photosynthetic membranes. *Int. J. Quantum Chem.* **17**, 587–93.

Ruff, I. (1965). The role of water in electron-transfer reactions: I. *J. Phys. Chem.* **69**, 3183–6.

Salemme, F. R. (1976). An hypothetical structure for an intermolecular electron transfer complex of cytochromes *c* and b_5. *J. molec. Biol.* **102**, 563–568.

Salemme, F. R. (1977). Structure and function of cytochromes *c*. *A. Rev. Biochem.* **46**, 299–329.

Salemme, F. R. (1979). Structure/function relationships in biological electron transport proteins. **T**, pp. 523–40.

Salerno, J. C., Harmon, H. J., Blum, H., Leigh, J. S. & Ohnishi, T. (1977). A transmembrane quinone pair in the succinate dehydrogenase–cytochrome *b* region. *FEES Lett.* **82**, 179–82.

Salerno, J. C. & Ohnishi, T. (1979). Electron transport in the succinate-ubiquinone segment of the respiratory chain. **T**, pp. 473–82.

Sarai, A. (1979). Energy-gap and temperature dependence of electron and excitation transfer in biological systems. *Chem. Phys. Lett.* **63**, 360–6.

Sarai, A. (1980). Possible role of protein in photosynthetic electron transfer. *Biochem. biophys. Acta* **589**, 71–83.

Sarai, A. & Kakitani, T. (1981). Radiationless transitions of molecules with large nuclear rearrangement. *Chem. Phys. Lett.* **77**, 427–432.

Sarna, T., Duleba, A., Korytowski, W. & Swartz, H. (1980). Interaction of melanin with oxygen. *Arch. Biochem. Biophys.* **200**, 140–8.

Sauer, K., Mathis, P., Acker, S. & Van Best, J. (1978). Electron acceptors associated with P-700 in triton solubilized photosystem I particles from spinach chloroplasts. *Biochim. biophys. Acta* **50**, 120–34.

Scalapino, D. J. & Marcus, S. M. (1967). Theory of inelastic electron–molecule interactions in tunnel junctions. *Phys. Rev. Lett.* **18**, 459–61.

Schenck, C. C., Blankenship, R. E. & Parson, W. W. (1982). Radical-pair decay kinetics, triplet yields and delayed fluorescence from bacterial reaction centers. *Biochim. Biophys. Acta* **680**, 44–59.

Schenck, C. C., Parson, W. W., Holten, D., Windsor, M. W. & Sarai, A. (1981). Temperature dependence of electron transfer between bacteriopheophytin and ubiquinone in protonated and deuterated reaction centers of *Rhodopseudomonas sphaeroides*. *Biophys. J.* **36**, 479–89.

Schneider, H., Lemasters, J. J., Hochli, M. & Hackenbrock, C. R. (1980). Liposome-mitochondrial inner membrane fusion. Lateral diffusion of integral electron transfer components. *J. Biol. Chem.* **255**, 3748–56.

Schrieffer, J. R. (1979*a*). Introductory comments. **T**, pp. 9–16.

Schrieffer, J. R. (1979*b*). Remarks. **T**, pp. 601–2.

Schrieffer, J. R., Scalapino, D. J. & Wilkins, J. W. (1963). Effective tunneling density of states in superconductors. *Phys. Rev. Lett.* **10**, 336–9.

Seibert, M. (1971). Spectral, kinetic and potentiometric studies of the laser-induced primary photochemical reactions in the photosynthetic bacterium, *Chromatium D.* Dissertation, Molecular Biology, University of Pennsylvania, Philadelphia.

Seibert, M. & DeVault, D. (1970). Relations between the laser-induced oxidations of the high and low potential cytochromes of *Chromatium D. Biochim. biophys. Acta* **205**, 220–31.

Seibert, M. & DeVault, D. (1971). Photosynthetic reaction center transients, P_{435} and P_{424}, in *Chromatium D. Biochim. biophys. Acta* **253**, 396–411.

Shapiro, S., Smith, P. H., Nicol, J., Miles, J. L. & Strong, P. F. (1962). Superconductivity and electron tunneling. *IBM J. Res. Develop.* **6**, 34–43.

Shuvalov, V. A., Ke, B. & Dolan, E. (1979). Kinetic and spectral properties of the intermediary electron acceptor A, in photosystem I. Subnanosecond spectroscopy. *FEBS Lett.* **100**, 5–8.

Siebrand, W. (1967). Radiationless transitions in polyatomic molecules. I. Calculations of Franck–Condon factors. *J. chem. Phys.* **46**, 440–7.

Slater, E. C. (1953). Biological oxidations. *A. Rev. Biochem.* **22**, 17–56.

Sonneveld, A., Duysens, L. N. M. & Moerdijk, A. (1980). Magnetic field-induced increase in chlorophyll *a* delayed fluorescence of photosystem II: A 100- to 200-ns component between 4.2 and 300 K, *Proc. natn. Acad. Sci. USA* **77**, 5889–93.

Sonneveld, A. (1981). Primary photochemistry and energy transfer in photosystems I and II of photosynthesis. Thesis, Rijksuniversiteit, Leiden.

Soules, T. F. & Duke, C. B. (1971). Resonant energy transfer between localized electronic states in a crystal *Phys. Rev.* B **3**, 262–74.

Sutin, N. (1962). Electron exchange reactions. *A. Rev. nucl. Sci.* **12**, 285–328.

Sutin, N. (1979). Electron transfer reactions of metal complexes in solution, **T**, pp. 201–24.

Swartz, H. M. & Dodd, N. J. F. (1981). The role of ascorbic acid on radical reactions *in vivo*. In *Oxygen and Oxy-radicals in Chemistry & Biology* (ed. M. A. J. Rodgers and E. L. Powers), pp. 161–8. New York: Academic Press.

Sybesma, C. (1969). Two light-induced reactions in photosynthetic bacteria. In *Progress in Photosynthesis Research* (ed. H. Metzner), pp. 1091–1100. Tübingen.

Szent-Györgyi, A. (1941). Towards a new biochemistry? *Science, NY* **93**, 609–11.

Szwarc, M. & Jagur-Grodzinski, J. (1974). Ions and ion pairs in electron transfer reactions of radical anions, carbanions and solvated electrons. In *Ions and Ion Paris in Organic Reactions* (ed. M. Szwarc), pp. 1–150. New York: Wiley.

Tafel, J. (1905). Über die Polarisation bei kathodischer Wasserstoffentwicklung. *Z. phys. Chem.* **50**, 641–712.

Takano, T., Kallai, O. B., Swanson, R. & Dickerson, R. E. (1973). The structure of ferrocytochrome *c* at 2.45 A resolution. *J. biol. Chem.* **248**, 5234–55.

Takuma, K. Sakamoto, T., Nagamura, T. & Matsuo, T. (1981). Novel properties of the self-assembling amphipathic viologen system. 1. A study of electron-exchange reactions in micellar systems. *J. Phys. Chem.* **85**, 619–21.

Takuma, K., Shuto, Y. & Matsuo, T. (1978). Studies on photoactivated electron-transport systems. II. Electron-transfer quenching of photoexcited ruthenium (II) complexes by various bipyridinium ions. *Chem. Lett.* (*Japan*), 983–6.

Tamura, N. Yamamoto, Y. & Nishimura, M. (1980). Effect of surface potential on P-700 reduction in chloroplasts. *Biochim. biophys. Acta* **592**, 536–45.

Taube, H. (1975). Intramolecular electron transfer. *Pure appl. Chem.* **44**, 25–42.

Taube, H. (1979). Experimental approaches to electronic coupling in metal ion redox systems. **T**, pp. 173–97.

Taylor, P. D. (1959). General discussion. *Disc. Faraday Soc.* **27**, 237–9.

Thornber, J. P. (1970). Photochemical reactions of purple bacteria as revealed by studies of three spectrally different carotenobacteriochlorophyll-protein complexes isolated from *Chromatium*, strain D. *Biochemistry* **9**, 2688–98.

Thurnauer, M. C., Katz, J. J. & Norris, J. R. (1975). The triplet state in bacterial photosynthesis: Possible mechanisms of the primary photo-act. *Proc. natn. Acad. Sci. USA* **72**, 3270–4.

Tiede, D. M., Leigh, J. S. & Dutton, P. L. (1978). Structural organization of the *Chromatium vinosum* reaction center associated *c*-cytochromes. *Biochim. biophys. Acta* **503**, 524–44.

Toyozawa, Y. (1967). Vibration-induced structures in the absorption spectra of localized electrons in solids. In *Dynamical Process in Solid State Optics* (ed. R. Kubo and H. Kamimura), pp. 90–115. New York: Benjamin.

Tric, C. (1971). Nonradiative transitions: A unified Green's function formalism. *J. chem. Phys.* **55**, 4303–9.

Trukhan, E. M. (1966). Determination of the mobility of free charged carriers in biological compounds. *Biophysics* **11**, 468–77. (Engl. Trans.: of *Biofizika* **11**, 412–19).

Ulstrup, J. (1973). Relationship between the energy of activation and the overall free energy of bridge-assisted electron transfer reactions in polar media. *Acta chem. scand.* **27**, 1067–72.

Ulstrup, J. (1977). Catalysis of the electrochemical reduction of molecular dioxygen by metal phthalocyanins. *J. electroanal. Chem.* **79**, 191–7.

Ulstrup, J. (1979). *Charge Transfer Processes in Condensed Media.* 419 pp. Berlin: Springer-Verlag. (No. 10 in series: *Lecture Notes in Chemistry.*)

Ulstrup, J. & Jortner, J. (1975). The effect of intramolecular quantum modes on free energy relationships for electron transfer reactions. *J. chem. Phys.* **63**, 4358–68.

Vanderkooi, J. M. (1976). Photoreduction of membrane bound cytochrome *c* by excited-state phenothiazine. *Biochem. biophys. Res. Commun.* **69**, 1043–9.

Vanderkooi, J. M. & Erecinska, M. (1975). Cytochrome *c* interaction with membranes. Absorption and emission spectra and binding characteristics of iron-free cytochrome *c*. *Eur. J. Biochem.* **60**, 199–207.

Vanderkooi, J. M. & Landsberg, R. (1977). Characterization of the cytochrome *c* binding site in mitochondria using metallocytochrome *c* derivatives. *Biophys. J.* **17**, 249a. Abstract F-POS-N3.

Van Duyne, R. P. & Fischer, S. F. (1974). A nonadiabatic description of electron transfer reactions involving large free energy changes. *Chem. Phy.* **5**, 183–197.

Van Gelder, B. F. & Beinert, H. (1969). Studies of the heme components of cytochrome *c* oxidase by EPR spectroscopy. *Biochim. biophys. Acta* **189**, 1–24.

Van Gelder, B. F., Orme-Johnson, W. H., Hansen, R. E. & Beinert, H. (1967). Electron paramagnetic resonance of heme at intermediate oxidation states of cytochrome *c* oxidase. *Proc. natn. Acad. Sci. USA* **58**, 1073–9.

Van Leeuwen, J. W. (1981). Pulse radiolysis studies of electron transfer processes in polar glasses. Thesis Rijksuniversiteit, Utrecht.

Vedeneyev, V. I., Gurvich, L. V., Kondrat'yev, V. N., Medvedev, V. A. & Frankevich, Ye. L. (1966). *Bond Energies Ionization Potentials and Electron Affinities.* London: Edward Arnold.

Vermeglio, A. & Mathis, P. (1973*a*). Photooxidation of cytochrome b-559 and the electron donors in chloroplast photosystem II. *Biochim. biophys. Acta* **292**, 763–71.

Vermeglio, A. & Mathis, P. (1973*b*). Photoreduction of C-550 and oxidation of cytochrome b-559 in chloroplasts. Dependence on the state of photosystem II. *Biochim. biophys. Acta* **314**, 57–65.

Vernon, L. P. & Shaw, E. R. (1971). Subchloroplast fragments: Triton X-100 method. In *Methods in Enzymology* **23** (ed. A. San Pietro), pp. 277–89. New York: Academic Press.

Visser, J. W. M., Rijgersberg, K. P. & Amesz, J. (1974). Light-induced reactions of ferredoxin and P700 at low temperatures. *Biochim. biophys. Acta* **368**, 235–46.

Visser, J. W. M., Rijgersberg, K. P. & Gast, P. (1977). Photo-oxidation of chlorophyll in spinach chloroplasts between 10 and 180 K. *Biochim. biophys. Acta* **460**, 36–46.

Voevodskii, V. V., Solodovnikov, S. P. & Chibrikin, V. M. (1959). Issledovanie spektrov elektronnovo paramagnitnovo resonansa (EPR) otritsatelnik ionov nekotorik aromaticheskik i heterotsiklicheskik soedinenii. *Dokl Acad. Nauk SSSR*, **129**, 1082–4.

Vorotyntsev, M. A. & Itskovich, E. M. (1980). Primary photosynthetic processes: the problem of rapid irreversible redistribution of electronic energy. *J. theor. Biol.* **86**, 223–36.

Vredenberg, W. J. & Duysens, L. N. M. (1961). Light-induced oxidation of cytochromes in photosynthetic bacteria between 20 and −170°. *Biochim. biophys. Acta* **79**, 456–63.

Waka, Y., Hamamoto, K. & Mataga, N. (1978). Pyrene-N,N-dimethylaniline heteroexcimer systems in aqueous micellar solutions. *Chem. Phys. Lett.* **53**, 242–6.

Walker, E. H. (1977). Quantum mechanical tunneling in synaptic and ephaptic transmission. *Int. J. Quantum Chem.* **11**, 103–27.

Warshel, A. (1979). Conversion of light energy to electrostatic energy in the proton pump of *Halobacterium halobium*. *Photochem. Photobiol.* **30**, 285–90.

Warshel, A. (1980). Role of the chlorophyll dimer in bacterial photosynthesis. *Proc. natn. Acad. Sci. USA* **77**, 3105–9.

Warshel, A. & Weiss, R. (1978). On the efficiency of electron transfer reactions in proteins. In *Frontiers of Biological Energetics* (ed. P. L. Dutton, J. Leigh and A. Scarpa), Vol. 1, pp. 30–6. New York: Academic Press.

Watson, G. N. (1945). *A Treatise on the Theory of Bessel Functions*. Cambridge: University Press; New York: Macmillan.

Wei, L. Y. (1967). Possibility of electron tunneling through a nerve membrane. *Bull. Math. Biophys.* **29**, 411–18.

Wei, L. Y. (1969). Molecular mechanisms of nerve excitation and conduction. *Bull. Math. Biophys.* **31**, 39–58.

Wei, L. Y. (1971). Quantum theory of nerve excitation. *Bull. Math. Biophys.* **33**, 187–94.

Weiss, J. (1954). On the theory of electron-transfer processes in aqueous solutions. *Proc. R. Soc. Lond.* A **222**, 128–41.

Weissman, S. I. (1958). Intramolecular electron exchange in anions of paracyclophanes. *J. Am. chem. Soc.* **80**, 6462–3.

Wentzel, G. (1926). Eine Verallgemeinerung der Quantenbedingungen für die Zwecke der Wellenmechanik. *Z. Phys.* **38**, 518–529.

Wever, R., Van Gelder, B. F. & DerVartanian, D. V. (1975). Biochemical and biophysical studies on cytochrome *c* oxidase. XX. Reaction with sulphide. *Biochim. biophys. Acta* **387**, 189–93.

Wherland, S. & Gray, H. B. (1976). Metalloprotein electron transfer reactions: Analysis of reactivity of horse heart cytochrome *c* with inorganic complexes. *Proc. natn. Acad. Sci. USA* **73**, 2950–4.

Wherland, S. & Pecht, I. (1978). Protein–protein electron transfer. A Marcus theory analysis of reactions between *c* type cytochromes and blue copper proteins. *Biochemistry* **17**, 2585–91.

Whitmarsh, J. & Cramer, W. A. (1977). Kinetics of the photoreduction of cytochrome *b*-559 by photosystem II in chloroplasts. *Biochim. biophys. Acta* **460**, 280–9.

Whitmarsh, J. & Cramer, W. A. (1979*a*). Cytochrome f function in photosynthetic electron transport. *Biophys. J.* **26**, 223–34.

Whitmarsh, J. & Cramer, W. A. (1979*b*). Theoretical time dependence of oxidation-reduction reactions in photosynthetic electron transport: Reduction of a linear chain by the plastoquinone pool. In *Photosynthesis and Nitrogen Fixation* (ed. A. San Pietro). Vol. 69 of *Methods in Enzymology*.

Wickramasinghe, N. C. (1974). Formaldehyde polymers in interstellar space. *Nature, Lond.* **252**, 462–3.

Wickramasinghe, N. C. & Santhanan, K. S. V. (1975). Reply to A. H. Fawcett (1975). *Nature, Lond.* **257**, 159.

Wiener, N. (1948). *Cybernetics*. New York: Wiley.

Wilkey, D. D. & Willard, J. E. (1976). Search for low activation energy abstraction of H from C-H bonds by D atoms in hydrocarbons at 77 K. *J. chem. Phys.* **64**, 3976–7.

Wilson, A. H. (1932). A note on the theory of rectification. *Proc. R. Soc. Lond.* A **136**, 487–98.

Wilson, D. F., Erecinska, M. & Owen, C. S. (1975). Some properties of the redox components of cytochrome *c* oxidase and their interactions. *Archs Biochem. Biophys.* **175**, 160–72.

Winfield, M. E. (1965). Electron transfer within and between haemoprotein molecules. *J. molec. Biol.* **12**, 600–11.

Witt, H. T., Moraw, R., Müller, A., Rumberg, B. & Zieger, G. (1960*a*). Kinetische Untersuchungen über die Primärvorgänge der Photosynthese. *Z. Electrochem.* **64**, 181–7.

Witt, H. T., Moraw, R., Müller, A., Rumberg, B. & Zieger, G. (1960*b*). Zwei chlorophyllreaktionen in vivo. *Z. phys. Chem. NF* **23**, 133–8.

Witt, H. T., Müller, A. & Rumberg, B. (1961). Oxidized cytochrome and chlorophyll C_2^+ in photosynthesis. *Nature, Lond.* **192**, 967–9.

Witt, K. & Wolff, C. H. (1970). Rise time of the absorption changes of chlorophyll-a, and carotenoids in photosynthesis. *Z. Naturf.* **25**b, 387–8.

Wolf, A. A. (1976). Experimental evidence for high-temperature organic fractional superconduction in cholates. *Physiol. Chem. & Phys.* **8**, 495–518.

Wolff, C. (1975). Formation of carotenoid triplets in photosynthesis bacteria. *Vth Int. Biophys. Congress*, Copenhagen, Abstr. p. 75.

Yamosa, S. (1978). On the basic equation for the equilibrium electronic states in polar solvents – broken symmetry. *J. phys. Soc. Japan* **44**, 602–10.

Yamosa, S. (1980). Theory of electron transfer in photosynthetic system. Preprint personally communicated Revised 1982.

Zener, C. (1932). Non-adiabatic crossing of energy levels. *Proc. R. Soc. Lond.* A **137**, 696–702.

Zener, C. (1934). A theory of the electrical breakdown of solid dielectrics. *Proc. R. Soc. Lond.* A **145**, 523–9.

Author index

Subject index